Global Construction Success

Global Construction Success

Charles O'Neil, FCIArb
Contract Dynamics Consulting

With contributions from 17 industry leaders

This edition first published 2019
© 2019 John Wiley & Sons Ltd

Registered Offices
John Wiley & Sons, Inc., 111 River Street, Hoboken, NJ 07030, USA
John Wiley & Sons Ltd, The Atrium, Southern Gate, Chichester, West Sussex, PO19 8SQ, UK

Editorial Office
9600 Garsington Road, Oxford, OX4 2DQ, UK

For details of our global editorial offices, customer services, and more information about Wiley products visit us at www.wiley.com.

Wiley also publishes its books in a variety of electronic formats and by print-on-demand. Some content that appears in standard print versions of this book may not be available in other formats.

Library of Congress Cataloging-in-Publication Data applied for

Hardback ISBN: 9781119440253

Cover Design: Wiley
Cover Images: © Rafik/iStock.com

Set in 10/12pt WarnockPro by SPi Global, Chennai, India

Printed in Singapore by C.O.S. Printers Pte Ltd

10 9 8 7 6 5 4 3 2 1

Endorsements for *"Global Construction Success"*

The authors of Global Construction Success are extremely grateful to have received the following endorsements from industry leaders in the U.K., U.S., South Africa, China, Malaysia, Singapore and Australia.

Lord Andrew Adonis   Chair of the UK National Infrastructure Commission 2017
Secretary of State for Transport 2009-10
Minister of State for Transport 2008-09

"Charles O'Neil and his co-authors have produced an impressive and important contribution to the construction industry that should be read by everyone involved in construction projects".

Nick Barrett   Editor of Construction Law Magazine, U.K.

"This book emphasising human factors and risk management in delivering successful construction projects comes at a potentially crucial turning point for the construction industry, with a new readiness to consider major changes to business models and processes evident following the Carillion collapse in the UK. The industry needs to read it".

Matthew Bell   Senior Lecturer and Co-Director of Studies, Construction Law,
Melbourne Law School, Australia

"Introducing this immensely useful book, Charles O'Neil writes that 'there is no better experience than learning the hard way'. This is true. Charles and his colleagues have generously shared their experience on construction projects around the world so that the rest of us can recognise and steer away from the commercial, technical and – especially – human factors which cause so many projects to founder".

Ann Bentley   Global Board Director, Rider Levett Bucknall
Member of the UK Construction Leadership Council

"A very well researched and easy to read book, which adds to the growing body of knowledge demonstrating that there is a better way of doing things in construction"

Gerhard Bester   MD of CAPIC, a South-African owned, specialist consulting services firm in the infrastructure development, construction and engineering industries.

"Africa's decision makers, both public and private, clients and contractors alike, should jump at the opportunity to acquire the benefit of hindsight from the industry in first world

countries – Africa generally follows their infrastructure delivery mechanisms, contracting regimes and unfortunately, consequential flaws… Africa has some additional variables to make things more challenging, but we cannot afford to ignore the wisdom and guidance on the way forward if we are to achieve "Global Construction Success" as presented by Charles O'Neil and his co-authors in this aptly named book!"

Chris Blythe   CEO, The Chartered Institute of Building, U.K.

"A great read with something for anyone wanting a successful construction industry. Construction is the art of getting ordinary people to do extraordinary things. Throughout the book, contributors show the best and worst behaviours that give the industry its extremes of reward and frustration. The wrong behaviours take the ordinary and produce the mediocre by repeating mistakes and not learning from them. Construction is too important as a driver of the global economy for the risks of failure to be as high as they are. This book offers an agenda for de-risking construction."

Dr. Fuyong Chen   Deputy Secretary-General of Beijing Arbitration Commission/ Beijing International Arbitration Centre; Vice-President of Asia Pacific Regional Arbitration Group

"Global Construction Success is a must-read for any decision makers in the construction industry. As a leading arbitration institution in China, Beijing Arbitration Commission/Beijing International Arbitration Center has accepted over 6000 construction cases since its establishment in 1995. These cases involved nearly any type of dispute that a project may have. However, all these disputes are definitely not inevitable. The recipe to avoid dispute is clearly included in the recommendations made by a stunning line-up of industry experts in this book".

John Dick   Partner, Dentons Rodyk & Davidson LLP, Singapore

"I have known Charles O'Neil for over 25 years, from his days running the Somerset Chancellor Court development in Ho Chi Minh City Vietnam. One thing about Charles that I have noted over the years is that when he expresses an opinion, it is worth listening to him. This work provides readers with a practical insight into understanding risks and how to manage them for successful project implementation. It highlights deficiencies in current infrastructure development frameworks, particularly relevant in the South East Asian markets, where Governments are relying more and more on PPP type frameworks for infrastructure development. Essential reading for construction success."

Mark Farmer   Author of The Construction Industry Review "Modernise or Die" 2016, U.K.

"I believe we stand at an unprecedented crossroads in the construction industry's evolution driven by a structural and long-term decline in skills and capability. This is no longer another false dawn driven by periodic discontent. The risks of continuing are now all consuming and include the increasingly destructive consequences of poor risk management and embedded conflict. The burgeoning technology led opportunity we are now presented with as our potential saviour will not be maximised though without embracing fundamentally different organisational, procurement and contractual models that drive process integration & common interest. This book is a very useful reference point using key lessons learned and pointing the way forward."

Professor Peter Hansford   Honorary Professor in the Bartlett School of Construction and Project Management, University College London, Former Chief Construction Adviser to the UK Government

Charles O'Neil and his contributors tackle the underlying issues in construction head on. This book sets out clear recommendations aimed at achieving tangible change. Building on his previous book, 'Human Dynamics in Construction Risk Management – the key to success or failure', O'Neil identifies human behaviour as a key ingredient to project success; this together with ensuring appropriate business models, embracing technology and of course adopting rigorous risk management – all subjects very close to my heart. I commend this book to students of project management, as well as to practitioners at all levels.

William A Marino   CEO, Star America Infrastructure Partners, LLC, New York, USA Chairman, Association for The Improvement of American Infrastructure (AIAI)

Charles O'Neil helps crystallize Public Private Partnerships as an important tool to be used in addressing America's crumbling infrastructure needs. Infrastructure in the U.S. has aged and, in many cases, has passed its useful life. The costs to revive this vital area, while addressing advances in technology, efficiency in process, and life cycle, are well documented by Mr O'Neil. Great work, well written.

Leighton O'Brien   Partner, Construction and Infrastructure Allens Linklaters Lawyers, Sydney, Australia

"O'Neil's book is a comprehensive review of the challenges and successes in the field. It will appeal to all participants in the construction industry as it is both approachable and authoritative.

The book presents a distillation of many years of experience and wisdom into an easily digestible format that utilises case-studies to reinforce concrete and practical lessons about successful construction projects.

Guest chapters by other notable figures within the industry provide a broad range of perspectives on a variety of topics internationally that gives the text added legitimacy and a global reach."

Datuk Sundra Rajoo   Director Asian International Arbitration Centre (AIAC), Kuala Lumpur, Malaysia Past President of the Chartered Institute of Arbitrators, U.K. (2016)

"This book is an exceptional collection of insight and wisdom from various experts across the global construction industry. It provides a 360-degree overview of the current state of international construction, including impacts of globalization, a detailed analysis of industry and regional trends in construction as well as the challenges faced by various sectors in the industry, making it relevant across the globe. This book is also written in simple and effective language, identifying the key areas of improvement within the industry and offering viable solutions for all stakeholders concerned. The author has also done

a remarkable job in structuring the book in such a way that makes it thorough and comprehensive, which is a boon for all of us in the industry. I believe this book will be a useful reference for all stakeholders concerned with navigating the emerging issues and challenges of risk management that plague this industry today"

Ian Rogers   Senior Legal Adviser, Arup, U.K.

"This hard-hitting collection of essays reveals the real problems in the construction industry, identifying not just the symptoms and how they might be treated, but also tackling the underlying causes. People and governance are top of the list and until these are addressed, change will be merely superficial. It is a vitally important contribution to the debate over the future of a key global industry".

Carol Stark   Managing Director, Aon Infrastructure Solutions, Chicago, USA

Global Construction Success provides an insider perspective on the construction industry as well as the PPP industry. The authors simplify the keys to a successful PPP procurement while challenging the industry to evolve. Their insights span the globe, highlighting the challenges a Special Purpose Companies face in implementing fulsome risk management programs when entering new geographies.

Don Ward   CEO of Constructing Excellence, U.K

"So many Governments and industry stakeholders around the world are anxious to see construction sector reform for major improvement in delivery. So why doesn't it happen faster? The insights in this book are hugely valuable to policy makers and industry leaders everywhere, with their focus on getting strong leadership and vision for projects, modernising the capability of people culture & behaviours in project teams, and aligning common processes and tools. Perhaps most crucial is the alignment of commercial arrangements throughout the supply chain."

Contents

Author's Notes

All opinions expressed in the contributors' chapters in this book are those of the authors and not of any organisation with which they may be associated.

All the other chapters that do not have a nominated author have been written by me, Charles O'Neil, and the opinions expressed in these chapters are mine alone and not of any organisation with which I am associated. The same applies to the quotations in boxes that have not been attributed to anyone in particular; some are commonly used in the industry and some are mine.

Notes on Project References

For obvious reasons we only refer to projects by name if the information quoted is publicly available on the internet from credible sources, such as from company announcements, court records, PPP portals, or approved in writing by the parties involved, and in each case we disclose the source.

Where the information is not publicly available, but we feel that important *'lessons learned'* have occurred we describe *'anonymous'* projects. With these references the key facts that caused the situation are based on a real project in each case and are accurately described without any embellishment, but the names of the parties and the location are not mentioned.

Acknowledgements

When I commenced putting this book together I sought advice from a number of my colleagues. This advice was so helpful that the conversation then led to *'why don't you write a chapter?'*

So, we have ended up with 17 contributing authors and I would like to thank all of you very much, because there is no doubt that your experience and insights have given the book a much broader and more interesting perspective. Additionally, you have provided many constructive comments along the way.

I would also like to thank all of my business colleagues and family members who have put in considerable effort in helping to get the book to the finish line, including my wife Andrea Slotnarin, Tony Blackler, Dr David Hancock, David Jardine, Damian Joy, Andy Matthews, Hugh O'Connor, Joanne O'Neil, Kenneth Reid, Frank Schramm, Thomas Schwenzer, Tim Sharpe, Gavin Stuart and Graham Whitson. No doubt there are others whose names I have overlooked who have been really helpful and I thank you equally.

When we completed the main draft of the manuscript we approached industry leaders in several countries to review our book and from this we have received their comments, which we greatly appreciate. On behalf of all of us who have been involved in writing the book, I would like to express our sincere thanks to all our reviewers for taking the time to read the book prior to publication and provide us with their generous endorsements. The purpose of this book is not to be critical of the construction industry for the sake of criticism. It is a call to action so that the industry takes its rightful place as a driver of the global economy for the benefit of all. I believe the comments we have received recognise the potential of the industry to achieve this.

With the production, the team at John Wiley & Sons Ltd have led the way, from initiating the book in the first place, and then with their excellent guidance and suggestions along the path to publication.

October 2018   *Charles O'Neil*

Biographies

Charles O'Neil DipArb FCIArb
Director, Contract Dynamics Consulting

Charles has over 40 years' experience in construction, steel fabrication, and asset management, predominantly in design and construction (D&C) and Public Private Partnership (PPP) projects – industrial; mining; grain terminals; motorways and bridges; hospitals, schools, prisons; shopping centres; multi-storeys, including offices, hotels, and apartments. With 'D&C' delivery, his career focus has always been on 'performance contracting', completing on-time and on-budget, based on very detailed planning, specifications, cost estimates, and contracts in order to minimise variations and cost increases for the clients and to avoid disputes.

Charles has worked in Australia, NZ, China, Vietnam, Malaysia, Singapore, Indonesia, Myanmar, UK, Europe, Chile, Canada, South Africa, and Rwanda, gaining an understanding of the different cultural and business practices in these countries. Since 2001 Charles has specialised in PPPs, with involvement in more than 60 projects in nine countries, in a variety of roles including structuring and bidding, design, construction, asset management, and in dispute resolution. He has undertaken appointments in several countries as an Arbitrator, Mediator, Dispute Board Member, Expert Determiner, Expert Witness, and Neutral Negotiator.

Currently, Charles' main activities are consulting on infrastructure and PPP risk management and undertaking infrastructure dispute resolution. He has written several publications on these topics and is a regular university guest lecturer and conference speaker on them.

Ian Williams MSc, CEng, FICE, FCIArb

Ian is a Chartered Engineer and a Fellow of the Institution of Civil Engineers and Institute of Arbitrators. He has over 30 years' experience in major engineering projects in the UK, Southern Africa, the Middle East, and South East Asia. Ian is currently working for the Supreme Committee responsible for transport operations for the 2022 FIFA World Cup Qatar™. Before joining the Supreme Committee, he was Head of Projects for the Government Olympic Executive on the London 2012 Olympic Games. Prior to the Olympics, he held key positions on the delivery of metros in London and Singapore, and airport infrastructure developments at Heathrow for Terminal 5 and the redevelopment of Terminals 3 and 4.

Judy Adams BSc, MRICS
Director and Global Head of Risk Management, Turner & Townsend

Judy joined Turner & Townsend in 1995 as Senior Cost Manager and was made Director in 2002.

She has worked in both the UK and overseas, for clients including BAA, The Wellcome Trust, Royal Bank of Scotland and the Department for International Development. She has also held senior management roles in Turner & Townsend's European and Asian businesses and was Global Operations Director from 2007 to 2011. Since 2011, Judy has performed the role of Global Head of Risk Management.

In addition to her current role, she is also one of Turner & Townsend's 96 global partners.

Professor Rudi Klein LLB, FCIOB, FRSA, Barrister, Cert Ed

Rudi Klein is CEO of the Specialist Engineering Contractors' (SEC) Group UK, an umbrella body representing the UK's premier trade associations in the specialist engineering sector. He is a barrister specialising in construction law and President of the New Engineering Contract (NEC) Users' Group. Rudi is acknowledged as the driving force behind the changes introduced over the years to improve payment performance in the UK construction industry. These have included the Construction Act and project bank accounts. He helped draft the Construction Contracts Act in Ireland, which was added to the Irish Statute Book in July 2013. Rudi is the author of two legal publications and countless legal articles in the construction press and academic journals. He is Visiting Professor of Construction Law at the University of Northumbria and Visiting Law Lecturer at the University of Stuttgart. In 2015 the Technology and Construction Court Solicitors' Association awarded him the Clare Edwards Award for his professional excellence and contribution in the field of construction law. In October 2016 Rudi was presented with a Lifetime Achievement Award by SELECT, the campaigning trade body for the electro-technical industry in Scotland. SEC Group was awarded the 2018 Gold Award at the H&V News Awards.

Stephen Woodward FAPM
Corporate Risk Management Consultant

Stephen has an interest in human behaviour (NLP Master Practitioner), principled negotiation and conflict avoidance. As a contractors' quantity surveyor (formerly FCIOB) he exploited his risk skills in claims generation. Influenced by construction management procurement he changed direction, becoming commercial director at a Japanese multinational contractor, pioneering early risk identification, fair risk sharing, and project mediation (CEDR Accredited Mediator). Subsequently working with funders, he moved project monitoring from a historical cost to a predictive risk service. For two decades he has been a construction risk advisor to project sponsors, funders and investors, and UK government offices.

Nigel Brindley MBA, IMC, BSc (Hons), CEng, MIMechE, MIFT
Director Infraquest Limited

Nigel advises investors in construction and infrastructure projects and businesses and is a non-executive director. He specialises in performance improvement and has advised a private equity firm in the acquisition of two distressed construction companies in 2018. Previously, Nigel led the asset management and investor reporting for Barclays Private Equity's infrastructure funds. He fulfilled a similar role at the investments division of Mowlem prior to its acquisition by Carillion. Nigel's background is in managing high profile construction projects (including BP Oil's new headquarters and the MI5's Thames House refurbishment). Nigel started his career at Arup where he was a design engineer on the iconic Lloyds Building in London.

David Somerset BSc LLB(Hons) FRICS FCIArb MEWI
Managing Director Somerset Consult UK

David is Managing Director of Somerset Consult (based in London), a claims, disputes and legal support services consultancy specialising in construction and engineering projects. David is a Chartered Surveyor and also holds a Law Degree.

David is a highly experienced expert witness, having been involved in a number of high profile disputes in the UK and internationally. The disputes include loss and expense claims, claims for rectification costs of defects and professional negligence claims. He is currently the quantum expert of a project assessing the construction costs in excess of US$1 billion.

David has written a number of articles relating to mechanical and electrical design responsibilities, claims from subcontractors and costing information. He is also advisor to a number of large organisations in the UK regarding procurement strategy and claims management. www.somersetconsult.com

Tony Llewellyn FRICS
Collaboration Director at Resolex Ltd., UK

Tony is a specialist in interpersonal dynamics and the effectiveness of project teams. Originally training as a surveyor, he has spent over 30 years working on major construction projects.

Tony has published two books on the topic of team coaching, *Performances Coaching for Complex Projects*, and *The Team Coaching Toolkit*. He is also a visiting lecturer at the University of Westminster where he runs a masters module on Developing Effective Project Teams.

Rob Horne CArb, FCIArb, LLM (Disp Res), FSALS, FFAVE (Master), LLB
Head of Construction Risk at Osborne Clarke LLP, London

Rob is a partner at Osborne Clarke, a leading international law firm. He received his Royal Charter as an Arbitrator over 10 years ago as one of the youngest ever Chartered Arbitrators. He is recognised by the legal directories for his innovation in finding solutions to complex project problems and as a leading expert in the NEC form of contract. He is the author of *The Expert Witness in Construction* published in 2013 and multiple guidance notes on adjudication published through the Adjudication Society and the Chartered Institute of Arbitrators. He is a committee member of the City of London Law Society Construction Group as well as being the chief examiner for the Royal Institution of Chartered Surveyors.

Rob's practice is largely based on major projects within the UK and internationally across the Middle East, Africa, and Europe, and even advising on projects in Australia, in which he is often brought into very early to help steer them through the commercial minefield. He has been involved in disputes up to a value of £250million all tending to be centred around complex engineering problems, significant delay and analysis of disruption caused as a result across transport, power, health commercial, residential, and many other project types.

Richard Bayfield BSc (Hons), MSc, CEng, CEnv, FICE, FCIArb
Construction Consultant, Risk Director and Adjudicator

www.richardbayfield.com
Richard began his career working for Costain on major projects including the Thames Barrier in 1976. He has had senior roles as a contractor, consultant and client. His recent experience includes several major capital projects including the Sainsbury Wellcome Centre at UCL, the new Lambeth Palace Library, the Gradel Quad at New College, Oxford and HS2. Some career highlights follow:
- Three times British Construction Industry Award (www.bcia.org.uk) winner as Client representative – 2016 – Sainsbury Wellcome Centre at UCL, 2013 – Bishop Edward King Chapel, Cuddesdon Oxford and 2002 – Honda New European Plant, Swindon.
- Member of Construction Minister's sounding board of 6 'key industry figures' appointed to assist the DTI/BEIS (2006/10).
- Member of London 2012 Olympics and Heathrow Terminal 5 Dispute Panels.
- Former Chairman and Honorary Member of Society of Construction Law (2006).
- Joint author of DTI/BEIS sponsored report analysing 40% construction savings made by Honda when developing its 600 acre site at Swindon (2001/2002).

Jon Lyle BBuild (Hons) UNSW,
Development Director, Grocon Australia

Jon Lyle is an experienced international development and construction executive with an extensive track record in emerging and developed markets, delivering major mixed use developments and masterplans. His project portfolio stretches from Australia and New Zealand, through Indonesia and Malaysia to the Middle East, in Bahrain, Abu Dhabi, Dubai, and Saudi Arabia.

With a project portfolio in excess of US$10 billion of delivered projects Jon's understanding of project completion and operations allows him to know the 'end game' and what needs to be done on day one to get there successfully.

Jon is a property development strategist who delivers developments using his extensive experience in both construction and development, allowing him to create high level development plans that incorporate all stakeholders and respective schedules of deliverables to ensure all interdependencies required are identified and complied with in accordance with the project plan. Jon is currently (2018) delivering the W Hotel for Marriott on Sydney Harbour.

Dr Robert Gaitskell QC BSc (Eng) FIET FIMechE FCIArb,
Keating Chambers, London, UK

Dr Gaitskell is the Chairman of the Dispute Board for the ITER nuclear fusion project in France, one of Europe's biggest power ventures. He is a Queen's Counsel, and acts regularly as a Dispute Board Member, Arbitrator, Mediator, Adjudicator and Expert Determinator in construction/engineering disputes worldwide. He is also a Chartered Engineer, a Fellow of the Institution of Engineering and Technology (FIET), a Fellow of the Institution of Mechanical Engineers (FIMechE), and a Fellow of the Chartered Institute of Arbitrators (FCIArb). As an engineer he was involved in the design of power stations and oil rigs. He is a former Vice-President of the IEE/IET (Europe's largest professional engineering body) and Senator of the Engineering Council and a part-time judge. Dr Gaitskell is the Editor of *The Keating Construction Dispute Resolution Handbook* 3rd edn, December 2016. His PhD, from King's College, London, concerned engineering standard form contracts. He is the Chairman of the IET/IMechE Joint Committee on Model Forms, which produces the MF/1-4 Suite of Contracts (used for power plants and other infrastructure projects).

John McArthur Dip Env Pl, BLA
Chairman, Kiewit Development Company, Toronto, Ontario, Canada and
Omaha, Nebraska, USA

John McArthur has 40 years' experience in the design, construction and investment business throughout North America. His education in urban planning and landscape architecture comprise his formal educational training and that, coupled with his years of experience in the design, construction, and infrastructure investment business throughout North America, form the basis of his understanding of risk management. During his career he has developed projects with one of Canada's largest real estate development companies throughout Canada. He also spent several years with a major Canadian construction company working on a variety of large building projects. Since 2003 he has been in the PPP investment business, initially for an international German construction/investment company and more recently with a large American construction/investment company. Over the course of a long career he has seen risk management develop to become the sophisticated process that it is today. Risk management is now widely employed by designers, subcontractors, owners, investors, rating agencies, insurance/bonding companies, and public agencies. He notes that in the past it was only given a minor level of recognition, but now it has become one of the trademarks of not only top tier companies, but also is a process utilised by most that are involved both directly and indirectly in the construction business.

Edward Moore MICW BSc

Edward has been Chief Executive at ResoLex UK since 2004. He is a specialist in embedding collaboration in teams and monitoring the effectiveness of team development. He created the RADAR methodology and toolset, which is a web based project horizon scanning and risk monitoring tool. The service has been developed to understand the impact of human dynamics on risk management and bridges the gap between forensic data and the perceptions of project stakeholders. It is now utilised on major projects to align team collaboration, behavioural risk and traditional risk management. Edward also designs and embeds alternate dispute prevention and resolution processes within project teams and commercial agreements.

Graham Thompson DipEng(Elec), LLB(Hons)
Graham is co-founder and CEO of Affinitext, a leading technology company which transforms the ability to quickly and confidently understand and manage complex contracts.

Graham has built an international reputation, first as an engineer and then as a key legal and commercial advisor on major infrastructure projects whilst a partner of Pinsent Masons and King & Wood Mallesons, with a lead role in developing successful collaborative contracting models, particularly project alliances.

This unique combination of international engineering, legal and commercial experience, coupled with a strong belief in the benefits of collaboration, led to a keen interest in developing smart technologies to transform the understanding and management of complex contracts.

His co-founding of Affinitext was a natural progression of this experience, which has led to the exponential global adoption of its disruptive Intelligent Document Format ('IDF') technology. www.affinitext.com

Professor Noha Saleeb PhD, MSc, BArch (Hons), FHEA

Dr Noha Saleeb is an Associate Professor in Construction & Creative Technologies, Middlesex University UK. She is Programme Leader for International MSc Building Information Modelling (BIM) Management, focusing on innovative digital technologies, enhancement processes of design, information, project and construction management. She is also Programme Leader for the BSc Architectural Technology programme. Her research interests are in architecture and IT, specialising in 3D digitisation and BIM, currently investigating implementation of Industry 4.0 technologies in design/construction, e.g. AI and Big Data analytics. Dr Saleeb is a practicing architect providing consultancy in BIM design, construction and onsite project coordination, and has achieved many awards and publications in journals, conferences, and books.

Stephen Warburton RIBA
Director, Design & Project Services Ltd.

Trained in Oxford and London, Stephen moved to Hong Kong in the mid 90s to pursue a career with development companies. After prior architectural internships in Washington DC and New York, Stephen was keen to work overseas again and became engaged on a diverse range of projects including luxury villas in Nha Trang, Vietnam, and Hong Kong's new international airport. On returning to the UK, he led Lend Lease's design teams on the BBC Mediacity Salford project. As a keen exponent of design in the built environment, he has also tutored students at the Oxford School of Architecture and has had projects exhibited at the Ashmolean Museum in Oxford, which were subsequently published in *A Treasury of Graphic Techniques* by Tom Porter. Stephen was a contributor to *Human Dynamics in Construction Risk Management* by Charles O'Neil in 2014.

Qualifications:

- BA (Hons) in Architecture. Oxford Brookes University
- Post-Graduate Diploma in Architecture. Oxford Brookes University
- MSc Construction Economics and Management. London South Bank University
- Chartered Member of the Royal Institute of British Architects.

David Richbell FCIArb (Mediation) 1943–2018.
David died on 25 September, 2018. Refer to our Tribute to David in the Preface

David had over 25 years mediating commercial disputes over numerous sectors. He had a particular interest in helping the mediator profession achieve recognition and he was responsible for revising the Chartered Institute of Arbitrators Mediation Rules and wrote their model procedure. He was a Board member of the Civil Mediation Council and was responsible for the introduction of the Registration scheme for Individual Mediators, Providers and Trainers.

David wrote the award-winning *Mediating Construction Disputes*, written for users, and the acclaimed *How to Master Commercial Mediation*, written for commercial mediators at every stage of their career.

Preface

Charles O'Neil, editor and author of 11/28 chapters.

Why Have I Written this Book?

I got a rude awakening to risk management in 1977 when I signed a fixed price contract with a two year programme for a coal handling plant for Gladstone Power Station in Queensland. During the next two years there was rampant wage inflation and we lost heavily, not the client. We were so keen to win the contract we excluded *'rise and fall'* from our offer. A hard lesson, but we survived and we certainly learnt from it!

In 2008, I was asked to investigate why a €300 million highway project cost more than €600 million and how two major companies in joint venture could get it so badly wrong? I quickly established that the cause was a series of shockingly bad decisions that overrode risk management checkpoints all the way up to Board level, all because a number of senior people *'really wanted to win the first project in that country'*. And in doing so they became completely blinded to or ignored the most obvious risk management *'stop-signs'*. It was a *'must win'* situation for all the wrong reasons. It only took me a short time to compare the costs per kilometre of similar projects in the data base of one of the companies, where I could see that the out-turn cost was always going to be around €600 million, even if they built the highway with their best efficiency. There were no excuses, just a lot of egg on face and embarrassing explanations to shareholders.

This investigation whetted my appetite, and out of interest I started looking into construction and engineering *'disaster'* projects worldwide and researching what went wrong with each of them. My initial objective was to prepare risk management information for internal use by my employer that summarised the most common causes and the lessons to be learnt. The first time round with this research, I identified 19 common causes of significant project failure, but with more investigation this expanded to 35 causes, as listed in Chapter 4. Of these, 20 occurred during the planning and pre-contract stage and 15 during the detailed design and construction stage

It became very apparent that in virtually every case the real cause was human behaviour at some level of the decision making or implementation stage. In most cases it was at a senior management level. I concluded that the real issue was that *'people are definitely the problem, not processes'* and the big question was how to reduce the potential for individuals to sidestep the system and cause all sorts of problems through their erroneous behaviour.

Since then I have taken an increasing interest in the *'human behaviour'* perspective in planning and construction, the effects that people may actually have on situations and how vital it is to understand this in establishing robust corporate and project management.

All this motivated me to write my first book *Human Dynamics in Construction Risk Management – the key to success or failure* that I published in 2014 and which chronicled my observations and conclusions, supported by contributions from leading experts in their field. The book was well received. The chapter on *'35 Common Causes of Project Failure'* attracted a lot of attention and we have updated and repeated it here.

Unfortunately since then there have been more high profile project failures and an apparent increase in contractor and supply chain companies collapsing or getting into serious financial difficulty.

My involvement in the construction industry has now lasted for more than four decades, in almost every aspect of it, starting as a welder in steel fabrication; then studying building and moving into site management, subcontracting and head contracting; into property development, Public Private Partnerships (PPPs) and asset management; to representing clients; and in dispute resolution. This has covered a wide range of civil and building projects and in several countries. I have enjoyed (almost) every minute of it and still do, just as much as at the beginning; some would say passionate about it.

The global construction industry is at a crossroads as we write this book in 2018, with some exciting new technical and digital developments, but at the same time the industry is being seriously hampered by the continuing use of outdated business models and terms of contract that need urgent revision.

Efficient and selective contractors are coping with this, but many are not, and they commonly blame low margins and onerous contracts and payment terms, although often these are just excuses for inefficient and incompetent management. However, it is a worldwide problem and the major issues are common almost everywhere. Construction has also been caught up in the push towards globalisation and today you find Chinese contractors in Africa, Koreans in the UK, Australians in the Middle East, and so on, but every second day you read about problem projects and construction companies in financial difficulty somewhere.

So, all of this has been the motivation for our book, which is the combined effort of our contributing authors who have all built their careers around the construction industry. We all share some very real concerns that the global industry is not changing and adapting many of its traditional practices and cultures rapidly enough to be a financially healthy and sustainable industry in this fast moving world, and by comparison with other industries.

My co-authors and I have written this book with the following purposes:

1) To advocate the urgent need for change to the outdated business models, processes, and cultures that are holding back the potential of construction contractors and their supply chain to be financially strong and profitable, and to do this by identifying and analysing the key areas of concern.
2) To use our combined experience to provide practical examples and make recommendations for improving management efficiency at the government, corporate and project levels in order to help the industry become more efficient and less risky for clients, contractors and investors.

It is a real truism of life that there is no better experience than learning the hard way. We all make mistakes and many of the most successful entrepreneurs and managers in a wide range of industries will tell you that those mistakes and the lessons they learnt from them were an invaluable part of their ultimate success. Once may be OK, but what is not acceptable is to make the same mistakes repeatedly, yet this is exactly what has been occurring in the construction industry for at least the last 40 years that this author has been involved with major projects and certainly it has been going on considerably longer.

Objectives

In line with our purposes, our first objective is to identify and analyse all the structural problems and key issues causing concern in the industry that affect the viability of contractors and the supply chain and which in turn affect their ability to deliver quality projects. These areas include:

- People, teamwork, and management competency
- Human behaviour and its impact on all areas
- Business models, forms, and terms of contract
- Corporate and project management processes
- Risk management.

We also highlight areas where the industry is making good progress with improvements to practices and processes.

We look at how people react in certain situations under pressure or for other extraneous reasons. The study of human factors is becoming widely accepted as a vital management component by many industries, in healthcare, with airline pilots, in factory production, and so on.

This includes all levels of personnel from Board members to site managers. We delve into the range of human behaviour, including defensiveness, aggression (bullies and sledgehammer managers), greed, ego driven behaviour, loss of perspective and reality, and bottom-drawer managers who hide or ignore things for different reasons.

From this, our next key objective is to make the *'lessons learnt'* available in an easy-to-read, concise and practical form, so that public and private sector managers can take them on board.

It is equally important to examine successful projects as well as those that get into serious trouble, so that effective comparisons can be made and the differences properly understood.

We review the key strategies and processes of corporate and project management, with comment on the good and bad practices and with recommendations for improving delivery efficiency and profitability.

Other key themes that we examine are communications, collaboration and relationship management, all of which rank highly in importance and they receive a lot of attention.

On an industry wide basis, there has definitely been too much emphasis on winning projects to chase revenue and not nearly enough on profitability, cash flow and prudent risk management. This 'must win' competitive approach is totally unacceptable when it results in ever-increasing borrowings and project and corporate losses. Managers need to remember that they are answerable to their shareholders and lenders and are responsible for the prudent use of the shareholder and loan capital.

In writing this book, the challenge we have faced has been to put our collective thoughts together in a frank and objective way that will encourage top level management to review the way in which they operate and ensure that they are running their construction companies like real businesses, and not just be 'monument' builders (a common term for revenue chasers).

At the same time, we believe regulators in each country need to really listen to the concerns of the industry and work collaboratively with industry organisations to implement the necessary changes that will overcome the major issues.

So what are the solutions? Certainly for public companies, senior management competence is paramount, with an end to cronyism and jobs for the boys. And accountability has to be reinforced by the regulators through serious penalties for cover-ups and false reporting.

What are employees and shareholders meant to think when it comes out that the pension fund has been raided to prop up the cash flow, or that the directors have given themselves pay rises or bonuses while the ship is sinking.

Given that most of the problems today have been occurring on and off for decades, it is clear that there need to be some radical changes to the industry culture. We are certainly not alone in thinking this, as someone is saying it publicly every second day. But major change is exceedingly difficult, with conscious and even sub-conscious resistance always a problem. It will only come about if industry leaders in both the public and private sectors agree and work together on implementing real change to the culture and business models.

My Journey from the Australian Bush to International Construction

My journey started 400 km west of Sydney in Australia; since then I have lived and worked in several countries and now reside in Germany, but I've been lucky enough to travel to various places around the world on a regular basis.

I grew up on my parent's sheep and cattle farm in central New South Wales. I started school riding a horse 5 km down the road to a one-teacher bush primary school with 16 pupils. My secondary education was in Sydney at a boarding school, where I passed the final year exams and left school at 16. My results were good enough to go to university, but this was never considered and I went back to the bush to learn about sheep, cattle and farming, often spending long days on horseback. 40 years later and after many winding roads I decided it was time to go to university and enrolled at Reading University in the UK to study Arbitration Law and Practice.

The year after leaving school I was fortunate to tour the world for six months playing cricket with an Australian club team; visiting Canada, the UK and some European cities, and then on to Pakistan, India, Sri Lanka, Singapore, and Indonesia. This opened my eyes to the fact that there was a lot more to the world than sitting on a horse chasing sheep and cattle. This experience unknowingly set the stage for my later international career.

I worked with my father on his farm for four years and then went 500 km north to manage a 7000 acre sheep and cattle property. I developed a passion for repairing machinery, welding and building steel yards and sheds on the farms. In 1969, we had an urgent need for a steel cattle crush, used for administering veterinary treatment to cattle in large numbers. Due to the expense, purchasing one was not an option. So a friend on another farm and I decided to design and build a steel one to fit our needs. We probably over-built them a bit as they are still working well 50 years later! On the back of this we commenced a manufacturing business and within 12 months had 25 employees. I bought out my partner and expanded the range of equipment and also moved into commercial steel fabrication. In 1972 and in 1974, I won awards for new equipment design at the Australian National Field Days, an annual show for farmers and stock breeders.

In 1974, I obtained my Construction Supervisor's certificate and Builder's Licence and our business entered into general construction. We expanded to 100 employees and completed our first significant project that year, the structural framework for a 30 000 m^2 shopping complex that we built in record time. We entered into a penalty/bonus contract with the client, with my team sharing 90% of an attractive performance bonus and the business keeping 10%.

This set the new direction for the business, which from then on concentrated on what we termed *'performance contracting'*: undertaking Design & Construct (D&C) contracts in Australia and New Zealand with demanding schedules and fixed lump sum contract prices. At the same time we undertook some property development with retail, commercial and residential projects. We gained experience in concept design, feasibility studies, financing and facilities management. As our *'own'* builder we quickly learnt the importance of the S-Curve when projects are being developed on loan finance. *(See Chapter 24 for an explanation of S-Curves.)*

In 1990–1991 Australia was in the grip of a recession, the construction industry was at a low ebb and it was hard to obtain new projects, let alone make a profit. South East Asia was booming so I toured several countries looking for opportunities and in early 1992 negotiated an agreement between Singaporean investors, as my backers, and a Vietnamese government company in Ho Chi Minh City to finance, design, construct and operate a 21 storey serviced apartment development. It was a 30 year contract that was effectively a PPP. This was one of the first joint ventures established after Vietnam opened its doors to foreign investment and it was not an easy road. The project was successfully completed in September 1996 and continues today as one of the most successful joint ventures in Vietnam.

This was the start of my international career in managing developments and construction projects. The high-rise project in Vietnam presented a huge learning curve for me in respect of understanding and managing a project involving multi-national participants, cross-border contracts and financing.

I lived in Ho Chi Minh City for five years and then in Kuala Lumpur for four years. My involvement in design and construction during this time included factories in Vietnam for foreign manufacturers, shopping centre and industrial park developments in Malaysia, and the infrastructure and utilities design for a new 500 acre township near Yangon in Myanmar, completed before the international sanctions were put in place. These activities introduced me to a broad range of interesting companies, design consultants, and government organisations in these countries and in Singapore and Hong Kong.

Looking back, the most significant experience that I gained was an understanding of the diverse cultural and business practices in the different Asian countries I worked in.

In 2001, I joined the German company Bilfinger Berger AG in the UK, in their PPP division. To me this was a great challenge because PPPs covered feasibility and financing, design and construction through to the long term services operations of the facilities. PPPs are partnering in the true sense and if you don't get it right then the financial and reputational losses can be heavy. The ingredients for successful PPPs are well covered in this book, but in short PPPs are all about people, and the projects rely heavily on the human dynamics working in a positive manner. Since 2001, I have had continued involvement on a wide range of PPPs.

There can be very real human reasons why a project must be completed on time. For example, take the new maternity hospital that was replacing three worn-out old hospitals in the UK. The National Health Service (NHS) and local authorities publicly announced the opening date four months ahead and at the same time sent out letters to 1000 pregnant mums advising them they could have their babies in the new hospital. One month before the opening date the builder announced the completion would be a month late, which was not acceptable. As the lead person, I was not going to face those pregnant mums! The hospital opened on time! My earlier experience in performance contracting certainly helped in that situation.

Acceleration is a fine line between achieving the target date and blowing the budget, but it does not have to be all about throwing money at it to fix it; in fact there is a much better way. If you lead from the front, motivate and encourage your management team, the builder and their subcontractors, it is not difficult to develop a team spirit whereby no one wants to be the one that lets the team down. People really will go that extra mile on that basis. Empathy, relationships and communications are all important.

During my time working with Bilfinger on PPPs, I became more and more interested in risk management and how vital it is to understand the human element. In 2009–2011, I was appointed director of asset management for the Bilfinger global infrastructure portfolio, overviewing design and construction and operational performance of a range of investments in several countries, including motorways and bridges, hospitals, schools, prisons, and government administration buildings. During this period, I led a team that restructured the global financial and operational reporting into a standardised real-time process and implemented stringent 'look-ahead' (early warning) risk management procedures.

My interest in dispute resolution goes back a long way because of my involvement in contracting. In 2001–2002, I studied arbitration law and practice at Reading University in the UK. The principles of arbitration led me to change my thinking about the way in which a problem should be approached. Arbitration teaches you to look at situations that are confused or have 'shades of grey' and to make them 'black and white' by unearthing and assessing the facts and not to be confused by emotional and self-interested interpretations, to view things impartially, how to read contracts and determine the rights and obligations of the parties, and so on. It can be recommended as great training at any stage in your career, even if you never go on to be an active arbitrator.

Who Should Read this Book and Why?

We hope that this book contains worthwhile material for the decision makers in the construction industry – directors, CEOs and senior managers of government procurement departments and contractors, and all stakeholders in projects from inception through design and construction to facility management and the end users. Our observations, analyses, and recommendations throughout the book apply equally to all these people, from planning authorities to the design team, the commercial managers and cost planners, bankers and lawyers; the complete supply chain. Last but by far not least are the people who will use the facilities and infrastructure, the asset managers and services industry. They are the users who inherit the assets and have to adopt, change and operate less efficiently if their functional needs are not met. They should be involved from the outset.

One of the key objectives of this book is to provide information that will be of benefit to industry participants in both developed and developing countries.

Conclusion

In putting this book together I have been privileged to have received generous and enthusiastic support from my co-authors, who are all experienced industry leaders in the UK, Europe, North America, and Australia. Their contributions clearly illustrate the movement in the industry towards greater efficiency and reduced risk. It is vital that head contractors, consultants, subcontractors, and suppliers remain viable and profitable in order to be able to deliver quality projects to their clients. It is a highly competitive industry, but one that is a key component of the economy in every country.

I wish to wholeheartedly thank my co-authors and John Wiley & Sons Limited for their time, contributions and collaboration in publishing this book.

Tribute to David Richbell who died on 25th September, 2018, aged 75.

David was a renowned mediator whose professional background was in construction and property and he specialised in disputes in these areas. He was also recognised as one of the most experienced and successful general commercial mediators in the U.K., Europe and internationally, with a particular ability to win the trust and confidence of the parties.

With an engaging and enthusiastic personality, David was a highly regarded trainer and mentor of commercial mediators, with affiliations to many prominent industry, academic and dispute resolution institutions around the world.

David wrote Chapter 25 for our book *"Managing and Resolving Conflict"*, but passed away during the preparation stages for publishing. We are honoured to include his words of wisdom, which shall be relevant forever.

1

Introduction

Ian Williams MSc, CEng, FICE, FCIArb

Former Head of Projects for the Government Olympic Executive, London 2012

1.1 Opening Remarks

Things do go wrong on construction projects – surprisingly often and for a range of remarkably similar reasons! These failures are not confined to one country or region or culture. They happen across all continents, and have done so for many years.

So what are these failings? Could they be related to construction being an industry that has always been notorious for operating on a high risk/low margin model? Might a relevant factor be the onerous payment and contractual terms that are often imposed on head contractors and on consultants by clients, or by head contractors on their sub-contractors and suppliers? Are companies bidding too aggressively and too often for work on the basis that this project is a 'must win' – either for the (short-term) prestige of tender success, or the necessity to win new work just to cover failings elsewhere in their business? Are individuals over-optimistic regarding outcomes and not evaluating the risks associated with time and cost in a robust, realistic, and pragmatic way?

Each new generation inevitably finds itself repeating the same old questions of 'why' the industry continues to be like this, and then 'how' and 'what' can we do to improve it for the future.

This book seeks to identify and understand the key structural 'why' questions related to the construction industry at this time, and then to make constructive arguments as to 'how' we can improve in the future, with emphasis on 'what' specific actions and focus could produce a long-term improvement in the industry's general health.

From the experience of the authors and their analyses of failures, four overarching themes emerge on a consistent basis:

1) *People and teamwork.* We all know that building a house or a wall needs to be done on sound foundations. The critical foundation theme is people and getting the team and its organisation right. This starts with commitment from people with the right behaviours and values working in committed, focused teams.
2) *The right framework (business models and form and terms of contracts).* Getting the right contractual environment is essential to allow competent people and teams to deliver successful project outcomes.
3) *Management of risk.* Managing risk is not only the responsibility of risk managers; it is everyone's job. Managing risk must be at the centre of each project and the process of

Global Construction Success, First Edition. Charles O'Neil.
© 2019 John Wiley & Sons Ltd. Published 2019 by John Wiley & Sons Ltd.

identifying, eliminating, and managing it must be a core management tool and value, not a bolt-on. Client and project sponsors need to consider the concept of who owns the risk and who is best placed to eliminate it and if that's not possible, how should it be managed to reduce its probability of occurring and its severity. Why is risk management no higher up the 'framework'? The simple answer to that is that it is equal, but without having the right people and contract arrangements in place the successful management of risk will not happen despite having the best risk management system in the world in place.

4) *Robust processes (corporate and project management).* Flowing from the contract and how it is set up are the robust processes that are essential to enable the project to be executed successfully. History shows that many projects fail at this hurdle.

The book is split into six sections. Section A sets the scene with an overview of the state of the industry. Sections B, C, D, and E cover the solutions we put forward. Finally, Section F summarises emerging conclusions. Each section is described in detail below.

1.2 Section A – The State of the Industry (Chapters 2–6)

This section of the book sets the scene with an overview of the current global state of our industry today. Chapter 2 takes a global overview from the start of the millennium. It looks at world cycles, industry and regional trends, and the good and bad news and their consequences. Following this in Chapter 3, Judy Adams, Director of Turner & Townsend, provides a really interesting overview of the issues faced by construction consultants in the international market place. Chapter 4 looks at the common causes of project failure and puts forward 35 common failure causes. In Chapter 5, Professor Rudi Klein addresses the issues of the use and abuse of construction supply chains and puts forward that the links for failure and success lay between communications, cash, people, and the product. Chapter 6 concludes this section by focusing on the UK construction crisis and the viability of tier-one contractors and supply chains, and shares lessons learned for improving the risk profile of infrastructure. At the end of this section the reader will be armed with an understanding of what the problems are that we face and will be equipped with a sound understanding of the issues that businesses currently face.

1.3 Section B – People and Teamwork (Chapters 7–11)

This section addresses the fundamental foundation block – people and how they are organised (teams and teamwork). These chapters provide solutions that address the foundation issues above all, which are all about people, their behaviours, values, and the organisational structures that they work in. In Chapter 7, we look at personal characteristics and corporate practices that can be obstacles to senior management and Board success. Chapter 8 focuses on structuring successful projects, with emphasis on the human element and effective teams. In Chapter 9, David Somerset, a respected dispute resolver, focuses on the criticality of understanding and managing difficult relations to achieve a successful win–win outcome. Tony Llewellyn in Chapter 10 puts forward that social intelligence is essential for project success. Finally, Chapter 11 brings this section of the book to a close, pulls all the strands together and focuses firmly on the human factors – getting the right people into the team. Without doubt, at the end of this section you will have grasped that it is people, people, people, and their behaviours and values, working in appropriate structures that set the right environment for success.

1.4   Section C – The Right Framework – Forms of Contract, Business Models, and Public Private Partnerships (Chapters 12–15)

Getting the right contractual environment is essential to allow competent people and teams to deliver successful project outcomes. This section looks at business models, the form and terms of contracts, and the need to create the right supply chain and commercial environment to allow competent people and effective teams, outlined in Section B, to work successfully – setting the rules and boundaries! In Chapter 12, Rob Horne, a 'seasoned' construction lawyer puts forward, with examples, that getting the contract right is a fundamental risk management tool. In Chapter 13, Richard Bayfield then discusses the New Engineering Contract (NEC) interface with early warning systems (EWSs) and collaboration. Chapter 14 puts forward Jon Lyle's personal international experience with development contracting. This section concludes with Chapter 15, which provides a general overview of Public Private Partnerships (PPPs) and makes recommendations for improvements to PPP contracts.

1.5   Section D – Management of Risk (Chapters 16–23)

Dr Robert Gaitskell QC opens this Section with Chapter 16 looking, not surprisingly, at human factors. Chapter 17 outlines effective risk management processes and John McArthur in Chapter 18 takes us though the relationship of success with risk management in North America. In Chapter 19 Edward Moore and Tony Llewellyn explain how an EWS provides the project community with simple but powerful early warning signals, which should be a standard part of all project management.

In Chapter 20, Rob Horne looks at advances in technology that aid in the management of construction risk. Graham Thomson in Chapter 21 focuses on intelligent document innovations that capture data and aid the processes of risk management and compliance. Professor Noha Saleeb in Chapter 22 looks at building information modelling (BIM) and highlights the organisational requirements that are needed for the successful implementation of BIM. Finally, in Chapter 23, Ian Williams and Stephen Warburton give an overview of specific projects where effective risk management has led to project success. The chapter also puts forward the lessons learnt from these projects, followed by a summary from Charles O'Neil on the importance of clear project ownership and leadership by senior management.

1.6   Section E – Robust Processes – Corporate and Project Management (Chapters 24–27)

Flowing from business models and the contract and how it is set up, the next key theme that needs to be addressed is the robust processes that are essential to enable the project to be executed successfully. Chapter 24 starts with addressing a fundamental project failure topic – time, i.e. failure to meet the schedule and the implications. Excellent advice is given to avoid the pitfalls that many have fallen into.

Notwithstanding having the right people, contract, and processes in place it is essential to have 'checks and balances' in place. It is healthy in a project to have challenging and open dialogue; nevertheless differences in opinion develop and can lead to disputes. In well thought out contracts a process on how to resolve disputes will exist. Chapter 25 looks at managing and resolving conflict and explains the real benefits and the success rate of mediation. Chapter 26 outlines essential areas of contract administration in respect of resolving disputes, in particular the benefits and risks of alternative dispute resolution methods. The final chapter in this section, Chapter 27, looks at the benefits of peer reviews and independent auditing. This is a very pertinent area that is being investigated in the wake of the Carillion corporate failure.

1.7   Section F – Emerging Conclusions (Chapter 28)

All the authors agree there is no 'magic formula' for the shift from 'business as usual' to achieving world class and successful project outcomes. Chapter 28 brings together the common findings and conclusions, puts forward recommendations, and challenges the reader to be bold in implementing change that will bring real efficiency, deliver world class projects and improve project and corporate profitability.

1.8   Final Note

Improvement in performance is never attained by 'business as usual'. If you undertake today's project in the same way you did yesterday, then tomorrow you will find yourself with yesterday's outcome!

The contributors are convinced that you will be better equipped to deal with tomorrow's projects if you consider and adopt some of the good practice ideas presented in this book. The extent of your success depends significantly upon whether you make the real commitment to learn, adapt, and change, and in just how far you are prepared to challenge 'old habits'

Tomorrow's outcomes are in the hands of all the many skilled professionals and leaders within our industry. Doing nothing is not really an option – good luck!

Section A – The State of the Industry

2

Global Overview of the Construction Industry

2.1   Introduction – Globalisation Impacts on Construction

In this chapter we review:

- World cycles
- Industry trends
- Regional trends
- Bad news and its consequences
- Good news – improvements in the right direction
- Summary and conclusions.

With the advent of globalisation construction has become quite a big 'export' market, e.g. Chinese, Japanese, and Korean construction companies have really spread their wings and readily compete in many developing nations with companies from the US, Canada, Brazil, UK, Europe, Australia, South Africa, etc. The 'client countries' are spread across many regions, in Africa, South and Central America, the Middle East, Central Asia and Asia Pacific. These international contractors are also providing healthy competition, or entering into joint venture contracts, in the home markets of their competitors, often introducing new methods.

2.2   Construction Industry Cycles

Construction industry cycles are often global and always interesting. The global financial crisis of 2007–2008 exposed the residential market, with notable boom–busts in North America, Ireland, Spain, Dubai, and Malaysia, amongst others. The outcome was that dozens of property developers collapsed, and their banks, contractors, and supply chains followed, or were badly hurt. In these countries, and others, the governments had to set up special bail-out entities to sort out the mess, which has taken a few years. Not so in Australia, where there has been an ongoing residential boom market since the mid-1990s.

We have also had mining, oil and gas, and heavy industry construction booms. Brazil, Chile, and Australia rode high on the mining boom from 2003 to 2014 with their massive export markets and even with the drop in prices that occurred mining is still big business for these countries.

Global Construction Success, First Edition. Charles O'Neil.
© 2019 John Wiley & Sons Ltd. Published 2019 by John Wiley & Sons Ltd.

The mining and oil and gas boom cycles have been good for engineering contractors, because these industries have been geared to EPC contracting (engineering, procurement, and construction), which in many cases amounted to 'cost-plus' contracts, with the clients having deep pockets when the oil, gas, and mineral prices were booming and they wanted their projects delivered as fast as possible. However, the demand tightened considerably when the drop came in commodity prices, with contract terms and pricing tightening at the same time.

At the time of publishing of this book, we are witnessing what we can call the 'infrastructure' cycle because there is an insatiable worldwide demand for new infrastructures of all types. It is predicted to keep growing at a fast pace, so as with most 'booms' people are rushing in to get part of the action,

2.3 Industry Trends – Business Models, Contract Types, Financing, Technology

A new trend in business models and contracting formats has emerged with the advent of 'soft loan' financing of projects by individual contractors in order to procure projects. Chinese and Japanese companies are prominent in this sort of project financing. This is in addition to the traditional financing in developing countries that is provided by the World Bank, the Asian Development Bank and the African Development Bank, amongst others, together with a relatively new entrant in this field, the Asian Infrastructure Investment Bank, an initiative of the Chinese Government that commenced operations in December 2015.

PPPs (Public Private Partnerships) add a further dimension by privately financing infrastructure developments, and they are spreading rapidly in developing nations as well as being well established in developed countries.

Another noticeable trend in the last two decades has been that many major construction companies have diversified into being asset owners and having their own services divisions to operate and maintain their asset portfolios, such as the case of ACS/Hochtief/CIMIC, Ferrovial, and Balfour Beatty in Europe. In South East Asia there are similar large conglomerates such as the Kuok Group and CK Hutchison Holdings, while the Macquarie Group has its tentacles all over the world. There are many other such conglomerates in India, China, Japan, North, and South America, and new ones are emerging throughout Africa.

With this vertical integration, these large groups finance, design, construct, own, and operate these assets.

Advances in technology are occurring rapidly and can greatly assist in improving efficiency and margins. In Chapter 20, Rob Horne provides us with a really interesting update and overview of Contech (technology to manage construction risk) and, in Chapters 21 and 22, Graham Thomson and Professor Noha Saleeb expand on IDF (intelligent document format) and BIM (building information modelling).

Use of these latest technologies is expected to grow rapidly and will surely define the difference between progressive and innovative managers and 'the old school'.

An increase in the sharing of risk on major projects has been another significant trend, such as with HS2 in the UK and the Gautrain High Speed Rail between Pretoria and Johannesburg in South Africa, completed in 2012. This is not only confined to high speed rail projects, but also to industrial facilities, cultural centres and multi-structure complexes such as in Doha for the 2022 FIFA World Cup.

There are some very large and very complex prestige projects being undertaken in several countries with adventurous geometric architecture, including cultural centres, scientific research facilities, space-age commercial premises, complex industrial plants, international airports, and so on.

One of the author's favourite new buildings is the Louis Vuitton Art Museum and Cultural Center in Paris that opened in 2014, an extraordinarily complex design by the architect Frank Gehry that was able to be built because of innovative digital architecture and engineering on a common web-hosted 3D digital model. The introduction of new technologies and skill sets that come with these international joint ventures is a very beneficial outcome. A good example of this was with the Mersey Gateway Bridge in the UK that opened for traffic in October 2017; three international contractors in a joint venture were appointed for the main contract, Kier from the UK, FPP from Spain and Samsung from Korea, and with specialist subcontractors for the cable-stay towers and the toll systems from Italy and France. A truly international and very successful project!

However, no matter how large, magnificent, and complex a new building, industrial or transport project may be, clients will always want to know the costs in advance and as accurately as possible. The rest of our book states how that should happen.

At the same time we must not lose sight of the fact that every-day basic construction is, and will continue to be, a very large component of the global construction industry, with housing and factories, schools and medical centres, and roads and bridges. The residential sector is actually the largest single sector in the industry.

The construction of these every-day basic necessities creates a huge employment base and is a fundamental component of the GDP in every country. The issues we are raising are essentially the same for all these sections of the industry

Just one interesting example of this employment base is that in Europe there are currently 180 000 people employed in the manufacture and placement of asphalt for roads, with many of the companies undertaking cross-border contracts. This is not surprising because there are five million kilometres of asphalt roads.

2.4   Regional Trends – Middle East, Asia Pacific, Africa, the Americas, UK and Europe

In the Middle East cultural practices have proven difficult for international contractors and construction consultants that have entered that market. Contract and payment terms are often ignored by government and private clients until final notices are issued. It is also common for contractors to have to put up performance bonds, of say 5–10%, that are all too easily redeemable for breaches of contract as interpreted by the client, which may include stopping work due to non-payment. Add retentions to the performance bonds and you are setting the scene for real financial difficulties on a project, especially if the client turns out to be a bad payer. On top of this you also may well be at the mercy of the local courts if you end up in litigation with contract disputes. You might ask who in their right mind would enter into a contract like that. Plenty of over-eager international contractors have in the past, but they are now becoming far more careful.

The Asia Pacific region, including India and China, has the largest volume of construction in the world.

China's infrastructure development has been really amazing in recent years. Just one recent high profile project is the Hong Kong–Zhuhai–Macau Bridge across the Pearl River Delta, which was opened in May 2018. Spanning 34 miles (55 km), this US$20 billion bridge is the world's longest sea-crossing bridge and is a monumental engineering achievement for China.

China also has the highest number of PPPs in the pipeline of any country and Indian PPPs are also gathering pace, with plans announced in 2018 for multi-billion US dollar rail and road PPPs.

When 'western' companies venture into these markets, which many have done successfully, it is really important that they understand the different cultures and practices between countries, because they vary considerably. Joint ventures, i.e. partnering, is a widely accepted business practice in this region on the basis that it should be a 'win–win' for both parties; the foreigner brings expertise and new techniques, and the local partner manages the domestic side of things and looks after the foreigner. Supply of funding is no longer an issue in most areas. Contractors, consultants, and suppliers who try to do it alone can fall into traps.

In Africa there is a huge future market, and major projects are being undertaken across the continent, but in some countries the levels of corruption are scaring away international investors and contractors who have to comply with the laws of their home countries, and non-compliance can attract heavy penalties. Development banks are having difficulty coping with the way their infrastructure funding is being misused.

In Central and South America there is continuing demand for mining, and oil and gas facilities, but apart from a few major projects such as the Panama Canal widening, there is not as large a demand for projects as in most other regions.

North America is experiencing a major push on infrastructures, with PPPs really taking off.

In the UK and Europe the industry continues to be very competitive, margins are very low, maintaining a positive cash flow is a real challenge and the supply chain is suffering at the hands of contractors. Structural changes in the industry are happening, but quite slowly. This is despite three UK government sponsored reports being published in the last two decades against a backdrop of deep concern in the industry and amongst its clients, of which the government is a major one. They have been the 1994 Latham Report, with one of its recommendations being fair payment; the 1998 Egan Report, which delved into all the major concerns and included the City's view that the construction industry was a poor investment; and the 2016 Farmer Report *Modernise or Die*, which basically covered the same ground as the two previous ones.

Professor Rudi Klein elaborates on these reports in Chapter 5.

However, here we are, 24 years after the Latham Report, with sectors of the UK industry still struggling with out-of-date contracting models and living from day to day on very tight margins and cash flows, with the supply chain also suffering as a consequence.

While there are clearly too many inefficiencies, and too much debt is being incurred to stay afloat, it is not a realistic solution to ask senior company management to say '*no, enough is enough and we won't bid any more on those terms'*. They are too busy running their companies.

Real structural change needs the joint efforts of industry groups in collaboration with the government. As we go to print it is encouraging to see that this might be finally happening in a meaningful way, with legislation being proposed to have retentions placed in trust accounts and 'bad payers' be excluded from government contracts. This is certainly a step in the right direction.

What changes have taken place since 1996? We have all seen headlines such as the one below in 2003 reporting on project failures:

> Never mind a West Coast upgrade – the route is off track
> It is difficult to tell which project, the new Scottish Parliament building or the upgrading of the West Coast Line, is more tragic in terms of delay, cost-overruns and bungled management.
>
> (*Scotsman Newspaper* July 2003)

Are these headlines a thing of the past? Unfortunately not!

In early 2018, the collapse of Carillion in the UK made global headlines. The extent of the fall-out and consequences extended beyond the UK to the Middle East, Canada, and Europe, and the publicity was widespread over the printed and broadcast media. The effects went far beyond the company; it affected ordinary people's lives ranging from the Carillion employees and its supply chain through to local communities depending on the assets Carillion were delivering, such as hospitals. It also extended to investors and in-house pension funds.

The extent of the consequences resulted in a parliamentary enquiry being launched in January 2018: '*Works and Pensions and BEIS committees into Carillion's collapse leaving a mountain of debt, potential job losses of thousands, a giant pension deficit and hundreds of millions of pounds of unfinished contracts with vast on-going costs to the UK taxpayer'*. The Committees investigated '*how a company that was signed off by KPMG as a going concern in spring 2017 could crash into liquidation with a reported £5 billion of liabilities and just £29 million left in cash less than a year later'*. www.parliament.uk – Work and Pensions Committee Carillion Joint Enquiry.

2.5 Bad News and Its Consequences

> Most other industries think that construction margins, the contract terms and the risks taken are ridiculous, but of course they demand highly competitive pricing when they want something built, so they won't be sympathetic.

And quite rightly so. When you look at the ongoing situation, with contractors in many countries continuing to get into financial trouble due to low margins and project losses, often their problems are self-inflicted by incompetent management entering into inappropriate contracts and then compounding the problem by managing them badly.

Carillion is not the only contractor to get into difficulties in 2017–2018, with several other large ones in Australia, New Zealand, Brazil, South Africa, and the UK either collapsing or announcing heavy project losses.

Just look at some more of the evidence:

- In New Zealand, significant contract losses were incurred by two of the nation's biggest construction companies in 2016–2017.
- In Australia, recent tier-two construction company failures have included Walton Construction, Cooper & Oxley, BuiltonCorp and back in 2012 Kell & Rigby, a 102 year old company that was a major player in government projects.
- In the UK, we have witnessed the failure of Carillion, Lakesmere, and several other sizeable companies; and in the previous 'cycle' that of Jarvis, Mowlem, and John Laing.
- In Germany, we have seen the demise of two major contractors, Philipp Holzmann AG in 2002 and Walter Bau AG in 2005 (which also took down Walter Group in Australia).

The above are just some random samples, but bad news seems endemic in the construction world and we are constantly reading about contractors getting into difficulties in the different regions, from Brazil to the Middle East, from South Africa to Asia.

Investors have become quite sceptical about over-optimistic pronouncements by companies in regard to their state of financial health. Typical of this was a statement in early 2018 by the CEO of a major contractor in which he referred to their improved underlying profit of £165 m and to their financial soundness with cautious optimism (quite correctly), but he did not highlight that it represented a margin of only 2% *before tax* on their annual turnover of £8.2bn. A couple of good project losses could wipe that out in the blink of an eye! The result was a considerable improvement on previous years, which just shows how fragile the margins are.

Is it any wonder that investors and shareholders get very concerned about the whole situation.

Case Study – in Australia, according to press reports, the NSW State Government is faced (in 2018–2019) with a delay of one year or more on the new AUD$2.2 bn. light rail system in Sydney, together with related variation cost claims of $1.2 bn. lodged by the head contractor. The delay and the variation claims are reportedly due to the government providing inadequate contractual information to the contractor in respect of the utilities to be moved and the scope of work involved, which has proved to be significantly greater than what was included in the contract. In other words the allegation appears to be that the government did not do its pre-contractual investigations and planning thoroughly enough and is now party to a contract that may have left it heavily exposed.

  If this is what has occurred, this would be a classic case of *'what happened to the detailed planning, scope of work identification, estimating and the risk management check processes, all this then flowing into the specifications and contract?'* A surprising outcome for a well-resourced government experienced in planning and implementing large scale infrastructure projects.

Why are corporate and project failures continuing to occur?
Where is the overall planning and implementation going wrong?

- Is it in the bigger picture: the concept, planning, the initial budgeting and cost estimates?
- Where is the overall project risk management breaking down?
- Do too many people in an organisation take the view that it is someone else's problem?

- Is it because people in the industry have become too specialised in their roles, do not always see the bigger picture, and do not communicate effectively between departments?
- Is there a need for greater multi-skilled training of project managers, e.g. competency in programming, estimating and cost planning, BIM, construction techniques, claims and progress payments, risk management for the whole project, communications and relationship management skills, etc.? This is how it used to be 30 or 40 years ago.
- Or is it because people are scared of speaking up against a dominant boss for fear of losing their job?
- Within companies, is it because CEO's and directors with big egos lose sense of reality and do not know when to stop empire building or excessive borrowing?

The ramifications of corporate and project failure are widespread and include:

- Significant delays in the delivery of the proposed services
- Reduced returns on investment
- Claims and disputes, which are costly in both time and money
- Substantial delays and losses for tenants and services contractors
- Breakdowns in business relationships
- Disruption to community services
- Political fallout and reputational damage
- High cost over-runs with heavy financial losses for investors, government clients, lenders, developers, contractors, insurers, and the supply chain; and at worst
- Bankruptcies for construction companies, subcontractors, and suppliers
- The administration of collapsed companies invariably soaks up any remaining cash at the expense of employees and creditors and is generally a bonanza for accountants and lawyers; the fees can be enormous.

2.6 The Good News – Significant Improvements in the Right Direction

It is important to recognise that there has been some real progress in the following areas:

- Forms of contract and their terms and conditions have improved substantially in the last couple of years, with the release of NEC4, the new FIDIC suite and other forms of contract that have been developed in countries to suit their particular conditions. Nevertheless it can still be argued that some of these contract formats do not yet offer enough protection to subcontractors and suppliers, or that the supply chain is still not able to enter into contracts on a level playing field and have to accept modified contracts, often with onerous terms, if they want the project. This means that legislative support is required in areas such as retentions, payment timing and security of payment.
- Technically, BIM and IDF are major steps in the right direction towards improving efficiency and are now being developed for all stages of a project. There is still some resistance to BIM, especially in relation to the supply chain, but this should diminish as people see the benefits. There are now many new buildings worldwide on which BIM and IDF have been used.

- As we publish this book in late 2018, Blockchain is being investigated to see how it might be of benefit to the construction industry to help improve efficiency, communications, security and payments, particularly with the supply chain, but there are significant challenges to overcome before it will be accepted and implemented. (See more in Chapter 20 on ConTech).
- Risk management processes have improved significantly, with early warning and real time control and reporting leading the way, but amazingly many contractors, government departments, and clients do not place nearly enough emphasis on risk management from the outset of a project and then wonder why when things go wrong. The new digital technologies can also make an important contribution to risk management.
- Health and safety – those of us who have been in the industry for more than 20 years have witnessed huge improvements in health and safety on sites in that time.
- Environment – in most countries the importance of the environment is now well recognised and must be accommodated in the concept and design in order to get planning approval. 'Green Star' ratings are also applicable in several countries.

In the UK the government has introduced some worthwhile initiatives that are a good start toward improving the efficiency of the construction industry including:

The Institution of Civil Engineers (ICE) and the Infrastructure Client Group (ICG) launched Project 13 in May 2018, supported by leading Government and private sector organisations. Project 13 is an industry-led initiative to improve the way high performance infrastructure is delivered, after organisations from all levels of the supply chain agreed that the infrastructure industry's current operating model was broken. Too often projects have been delivered over budget, past deadline and below par.

The Major Projects Leadership Academy (MPLA) for senior government managers, which has three simply stated objectives:

- Address common causes of project failure
- Significantly improve the quality of management amongst the leaders of critical projects *(including organisations, legal, and financial advisers)*
- Share learning across government.

(Note: We believe senior corporate management should put a lot more time and effort into this sort of approach).

At the time of publishing this book the UK government is proposing legislation to provide security for retentions and this will be really welcomed by the broader supply chain.

In June 2018 Network Rail took the lead and pledged to reform prompt payment practices and to abolish the use of retentions in its contracts. This is a very positive step in the right direction.

In 2017 the UK Government adopted the principal of collaborative working and published ISO 4401 (based on BS11000), entitled 'Collaborative business relationship management systems – requirements and framework', which includes weighted tender analysis towards non-confrontational company policies.

In regard to terms of payment and the maintenance of healthy cash flows, it is certainly worthwhile to keep a close eye on what is happening in other countries, e.g. in Australia the Security of Payment legislation that has been in place in all states and territories for some years that allows for the rapid determination of disputed progress claims under building contracts and contracts for the supply of goods or services in the building industry. This legislation was designed specifically to ensure continuing cash flow to these businesses.

There is now a good range of different contracting formats available so it should be possible for contractors and clients, including government procurement agencies, to negotiate terms in an appropriate form of contract that will safeguard the interests of both parties, provided they both accept the importance of collaboration and then manage their responsibilities efficiently.

These formats include Alliancing; ECI (early contractor involvement); EPC (engineering, procurement, and construction); IPI (integrated project insurance); penalty bonus (as distinct from liquidated damages and generally related to the completion time). These formats are supported by new forms of contract documents such as NEC4 and the FIDIC Suite, which can be used in the traditional form or amended to fit one of the above 'share the gain and pain' formats.

The use of these contracting business models is becoming more widespread as clients in both the public and private sectors recognise that it is in their best interest if they enter into construction contracts that are fair and reasonable to both sides, and which contain terms that effectively acknowledge that contractors and their supply chain need to be financially sound and must have positive cash flows.

2.7 Summary and Conclusions

The issues raised in this chapter highlight the need to have a healthy and robust global construction industry because funding institutions, trading banks and clients certainly do not want to be picking up the pieces and unexpected additional costs when a contractor goes bust or terminates a contract.

As previously stated, our primary focus in this book is on the qualifications, performance and accountability of CEOs, directors and senior managers in both the public and private sectors.

The reason for this is obvious. It is because 'the buck stops there' when it comes to running a successful and financially sound contracting or supply business, and for government and private clients in managing their end efficiently in the creation of successful projects.

This senior management group control and direct policies and strategies, finance, risk management, processes, personnel behaviours and cultures. So their skills and competence are paramount, because everything else flows from them.

With this in mind, it is necessary to ask some searching questions:

- Are too many directors and senior managers still embedded too much in the traditional management and cultural practices of the industry? Is training not keeping up with the times?
- Why aren't they adopting a much more hard-headed and professional business approach, including far more effective risk management, and stop their 'must win' chase for projects at any price?
- Are clients the problem with the continuance of the traditional contracting models because they think they have the advantage, to the extent that contractors are finding it difficult if not impossible to negotiate more reasonable contracts?

- <u>Are managers proactively managing human behaviour</u> to ensure that it is not causing problems in the different areas? There is no doubt that it can impact negatively on corporate management and project delivery in many ways.

Managers also have to ask themselves *'have their risk management processes adapted sufficiently to safeguard their interests'*; probably not in many situations.

When you look at the projects that we have reviewed in Chapter 4 and the common causes that we have identified, it is clear that the planning and risk management processes have failed badly in those examples. And the projects quoted were all in countries regarded as having robust checks and balances.

It is a real concern that these project disasters and corporate collapses have continued to occur all over the world on a regular basis and we want to get to the bottom of why this is so when we have advanced management training, many efficient risk management systems available, and much more awareness of the importance of communications and collaboration.

Even a big balance sheet won't help companies if they adopt the wrong strategies and/or employ incompetent new management.

<u>Investors, lenders, and other stakeholders are fed up with hearing all the excuses and promises</u> of CEOs that *'We have put into place rigorous processes to ensure a more disciplined approach towards project selection'*, blah, blah, blah (a recent comment by the CEO of a large contractor when announcing a big write-off). We have been hearing this in much the same terms on a regular basis for the last 40 years, but the buck does stop at the top and these processes should have been in place long before.

Investors are also fed up with construction companies that embark on over-ambitious expansion strategies, increasing their debt to prop up businesses in which they have no expertise or into new countries where they have no idea how to do business, especially if they attempt this without entering into a joint venture with an experienced local partner.

So beware of contractors whose debts are continually rising as they over-borrow to support their chase for low margin contracts and expansion plans, <u>yet do not invest anywhere near sufficiently in training, technology, and R & D because there is no cash left for this</u>; again just further examples of CEO and Board incompetence.

<u>It is important to understand the different cultural practices between countries.</u>

Many prudent international operators enter into a joint venture with a credible, successful, and influential local partner in order to try and safeguard their position. Even so, it can take two years or more to really understand how things work, based on the author's experience of working in several countries.

Some contractors and suppliers who have tried to go it alone in new countries have compounded their risks by entering into inadequate contracts, with insufficient due diligence and risk assessment. Even worse, some of them then proceed to appoint inexperienced 'expat' managers who run these contracts inefficiently, with poor administration of such things as variations, extensions of time (EOT's) and loss and expense claims. This in turn loses the respect of their clients, who then see it as an opportunity to take further advantage of them.

Is this scenario being unfair to contractors? Absolutely not, the author has witnessed this sort of bad management in several countries.

The safest way in a new region, as with any contract, is to spend the necessary extra time on detailed planning, precise specifications and on ensuring that the terms of contract and the payment procedures in particular are clear and unambiguous, fair and reasonable, and clearly understood by both parties when the contract is signed. It is not all that difficult for experienced lawyers and commercial managers to write contract terms that include the necessary safeguards.

If foreign clients will not agree to fair and reasonable contract terms and payment conditions, then you are better off without the business. Builders are builders, not bankers!

It should also be remembered that in every country it is important in negotiations to be respectful and polite, and to be careful not to cause loss of dignity or embarrassment, commonly referred to as 'loss of face'.

It seems to be taking a long time for the above messages to sink in. Clearly, new procurement methods and business models will only emerge if there is a concerted effort by industry institutions representing contractors and the supply chain working in conjunction with governments.

The opposite side of this debate on the financial health of the industry is to ask why should clients be able to demand performance guarantees and retentions, often followed by the arbitrary withholding of payment for concocted reasons, but not put up guarantees of their own in respect of milestone progress payments and final completion payments.

Contractors are not banks and they should be entitled to negotiate payment terms and guarantees that protect their financial position provided they perform the contract efficiently. The old idiom '*a fair days pay for a fair days work*' should always be applicable!

And clients cannot complain if they take on contractual responsibilities for which they are not best positioned or equipped to deliver, or do not clearly define the scope of the contract, or badly underestimate the costs in the first place.

Clients, including government authorities, have an equal responsibility to be competent and professional, and this does apply with the majority of clients. Those that are not have to bear the extra costs and a lot of publicity when their projects run off the rails.

On the positive side, it should be noted that there are efficient contractors in every country who are operating profitably in the current regime, but in any over-competitive market they really have to maintain that efficiency to survive, and invest in training, technology, and R&D.

The future is bright for construction contractors that are managed really professionally at all levels, but not so for 'old style' companies whose managers persist with traditional and unprofitable methods and who do not embrace change.

3

Construction Consultants in the Global Market Place

Judy Adams

Director and Global Head of Risk Management, Turner & Townsend

3.1 Introduction

This section provides an overview of the macro-risks facing global construction consultants currently. A number of these risks are likely to be the same for consultants and contractors alike. However, when delivering services globally, the risks are often magnified and it is harder to manage risk; as all regions the construction consultant operates in will have differing levels of maturity, the risks they face will be varied and the impact different.

3.2 Political Risk

One of the most significant changes in 2018 is an increased risk around changing political situations worldwide. Political crises cause instability, lack of confidence and sometimes safety fears, impacting investment and economic growth.

Public sector investment commonly reduces in the run up to national elections and erodes private sector investment confidence. Potential changes in policy post-election can also impact the timing and size of investment. This is amplified in the case of the uncertainty created by Brexit, which is impacting UK real estate values and construction volume, resulting in an exceedingly competitive market for construction consultants.

Economic sanctions increasingly appear to be a tool used by governments. This year (2018) has seen changing sanctions across a number of jurisdictions, including an embargo in Qatar, increased sanctions against Russia, and the introduction of potential counter sanctions by Russia.

Remaining compliant is difficult where the issues are varying, complex, difficult to research, and changing. When considering work in a sanctioned country, the construction consultant should assess each opportunity in a holistic manner, considering any knock-on effects to other regions in which they operate, e.g. in the past, even though some sanctions against Iran were relaxed, there was still anti-Iran sentiment in the US against companies doing work there because of its human rights record.

3.3 Regional/Cultural Differences

When operating as a global entity, there is obviously a requirement to comply with legislation in every jurisdiction. Increased business regulation, extra-territorial legislation and greater penalties for non-compliance have changed the global landscape. This can be particularly challenging when the consultancy is relatively immature, but operating in new jurisdictions.

Some areas that the global consultancy should be aware of are detailed below.

- Operating globally introduces challenges when setting up in new countries, including trade registrations, visa quotas, financial working requirements, employment law, the need to appoint sponsors or to involve third parties in company set ups or joint ventures to deliver and mitigate risk, as just some examples. Often, there are also financial regulations associated with repatriating money from the countries operated in to the 'centre' that need to be considered, along with the impact on exchange rates when reporting results.
- Construction is considered a high risk industry from a bribery and corruption perspective. With the introduction of pan-global regulation, such as the UK Bribery Act and the Foreign Corruption Practices Act 1977, and local legislation becoming more onerous, there is more at stake for construction consultancies if they get this wrong. Penalties/fines, prison sentences, and removal from Government tender lists are just a few of the risks. Where it does go wrong, Sweets in the UK being the most notable example to date relating to a construction consultancy, it can be significant – they were fined £2.3 m for failing to prevent an act of bribery, contrary to Section 7(1)(b) of the Bribery Act 2010 and £100 k by the RICS for bringing the profession into disrepute.

    While mandating compliance and promoting a culture of adherence to corporate governance will help reduce the risk, in many jurisdictions this may conflict with local customs and practice. There is therefore a need for the construction consultant to regularly assess areas of risk and put in place appropriate levels of assurance. As an example, in Russia, 'a three-person check' is often adopted. The theory being it's harder to bribe three people than two. An example of when this would be utilised is when undertaking peer reviews of tender recommendations.
- Tax authorities now also appear to be taking a more aggressive approach to cross border trading issues. Where services are delivered across multiple jurisdictions, the global consultancy should take external compliance advice and make sure their staff are aware of the issues and briefed accordingly.
- Because of the different legal systems across the world, the financial impact of an error (if and when it occurs), may differ substantially. The legal systems in the US, Canada, UK and Australia are probably the most litigious. There may also be legislation in play that introduces additional liabilities beyond normal contractual terms, e.g. the Australian Consumer Law, which the global consultant should be aware of and understand how to mitigate.

A number of these challenges will be exacerbated where the global consultant is delivering services in a location where they are not resident, and therefore, do not fully understand the issues facing them.

3.4   Payment or Fee Recovery

Delayed payment and/or non-recovery are risks in all businesses. However, over the last two–three years, there have been more examples of clients wishing to improve their cash flow at the expense of their consultants by delaying payment beyond 30 days, and where the consultant is acting as a sub-consultant, to only pay when paid. As a result, construction consultants are seeing increased collection risk through delayed payment and default, with a consequent impact on cash flow.

Ensuring that there is the requisite signed contract or purchase order in place won't wholly protect the consultant in getting paid, as in many jurisdictions the client will often cite 'poor performance' as a reason for non-payment; then it is up to the global consultant to try and recover their outstanding fees, where they may be less familiar with the legal framework, and where the level of fees may not warrant recourse to law.

It is also important for the construction consultant to be aware that in some instances non-payment may not be a reason to suspend or terminate their services, meaning that they may incur substantial costs without recovery.

Where there is a perceived risk of non-payment, then advance payments at the beginning of the project or requiring payment before the final issue of deliverables may help.

3.5   Localisation

As a result of growing national legislation and client requirements to improve the employment prospects of their nationals, professional consultancies, which have for many years relied on expat staff to deliver their services, need to localise their workforce more. Global consultancies need to address the specific issues in each region as part of an overall diversity and inclusion strategy.

Localisation does potentially cause risk in the short term due to a lack of skill and capability, but the benefits of this strategy include strengthening knowledge and insights into local markets, and maximising opportunities for growth. For global consultancies, this strategy will also help make them more competitive in the local market, without losing the other benefits they bring.

3.6   Failure to Attract or Retain Skilled People

Capacity, skills shortages, and strong competition in the industry makes resourcing a key risk for the construction professional, irrespective of whether they are working solely in the local market or as a global company. Brexit, legislative changes and government sentiment are also impacting resourcing models, reducing the already limited pool of people available.

For a global construction consultancy, the increasing size, complexity, and nature of its work also brings a requirement for higher calibre people. Indeed, some of the services provided are not recognised in some countries, making them reliant on expat staff to deliver, and where this resource is not present locally, language proficiency, visa restrictions, quotas, and other local legislation requirements make it harder to find the right resources to deliver.

In the medium to long term, there is a need to identify and train people locally, but this will continue to be a risk until resourcing models mature.

3.7 Contractual Terms and Conditions

The increasing trend for clients proposing more onerous contractual terms and/or looking to transfer more risk to their consultant continues. The biggest changes in the last few years appear to be around the need to accept unlimited liability and an increasing number of indemnities and warranties, which have longer term liabilities attached.

Predominantly in the Middle East, there is also a requirement to provide compliant tenders, coupled with a tender bond. Where the terms and conditions are particularly onerous, there is a danger that by qualifying their Bid, a consultant is either not considered for the work at all, with their Bid returned unopened, or for their bond to be called, should they be the preferred consultant, and then the consultant will be unable to negotiate better terms, resulting in them withdrawing from the process.

Liquidated and ascertained damages (LAD's), once the preserve of contractors only, are now finding their way into consultant contracts more often. That said, in certain jurisdictions, Russia as an example, the inclusion of LADs actually helps contain risk.

Consultants are also seeing more requests from clients to allow the information in their reports to be relied on by third parties, including external investors. As a result, there are more instances of consultants being pursued by the third parties for perceived errors in their information. Where possible, these types of request should be refused.

Onerous terms and conditions should be considered carefully as part of a Bid/No Bid process by the consultant, along with other risk factors, such as evaluating what work will be done in-house/sublet to others, the nature of the services to be performed, etc. and the risk fully understood. Where significant risk exists, or remains after negotiation with the client (and it is now more common for clients to refuse to negotiate), these opportunities should be declined.

3.8 Ability to Deliver Across Major Projects/Programmes

As a global consultancy, clients expect to be provided with a high quality, consistent service, no matter where the services are being delivered from. This creates operational challenges and reputational risk through reduced service standards and client satisfaction.

Consistency is particularly difficult given the differing operating models that may be in play, the experience of local staff, and the maturity of the construction supply chain as a whole.

3.9 Cyber Security

Cyber security is a high profile risk in every industry and no one is immune from the potential consequences. The entire threat landscape has significantly changed in the last

12–18 months and continues to change. Attacks are now well co-ordinated and very sophisticated, and there are more nation sponsored attacks, aiming to cause maximum damage, rather than to extort money.

Consultancies by their very nature rely heavily on IT systems to deliver their services. Their services also rely on analysing client data and other internal data, such as benchmarking information, etc. At its very worst, it is possible that a consultant would not be able to service clients at all. Data loss or an inability to access systems would also be a high risk.

Prevention is now unlikely to be an effective strategy. Therefore, more consultancies are placing greater importance on business continuity plans, retainers with business support companies and holding a copy of their IT system DNA in a secure location to allow an easier and prioritised system re-build in a worst-case scenario.

3.10   Contractor Failure

As economic pressures and increased competition to win work continues, the contracting supply chain is being impacted. As a result, contractors continue to be selective when tendering for work, resulting in difficulties in achieving full tender lists, and increased construction costs and delays affecting the construction consultant's ability to bring projects within their client's budget and programme.

These factors, coupled with often adversarial relationships, are increasing the pressures experienced in managing client projects.

In worst-case scenarios, contractors are going into receivership. When this happens, and there may have been prior problems with the contractor, it is more likely that the client will pursue the construction consultant as the 'last man standing'. Where the construction consultant is a large, well known company, they are likely to be the primary focus as the company with the most insurance.

Through working with the client and the supply chain to understand these risks and how to mitigate them, the construction consultant will be able to show value and reduce potential risk.

3.11   Design Liability

A major consideration for a global design consultant, operating globally, is around whether they are delivering the work and travelling or whether they have local capability or employ a subconsultant locally to support. Both areas have risks.

Where designing in a location different from the end location, the most obvious risk is associated with design standards, and it is possible for design consultants to 'over-design' or use incorrect codes. Where employing a subconsultant locally, the main areas of risk are associated with supply chain issues, particularly if the relationship is new, and design responsibility split, which may cause difficulties if there is a design error. In some instances, the local consultant may end up with a better relationship with the client representative, and further work may be awarded to the subconsultant directly.

4

Common Causes of Project Failure

4.1 Introduction

> Construction projects that turn into disasters have a very real similarity with aircraft crashes, in that 99% of the time the cause is human error

When we talk about *'project failure'* we first need to clarify just what that means, keeping in mind that 'failure' is likely to have a different meaning for different people, depending on their perspective. It can relate to one of the objectives of any party involved in a project not being met. This may include cost, time, client expectation, shareholder or boardroom expectation, safety targets, functionality or fit for purpose, public or political expectations and perceptions, etc.

While this chapter is essentially about individual projects, it should be kept in mind that large projects that are not efficiently managed can have the potential to put a contractor out of business. Signing the contract for a high profile construction project can be dangerous if all of the planning, design, specifications and contract terms are not properly in place and still have lots of grey areas or black holes.

In the following pages we have identified 35 common causes of project failure: 20 arising from actions taken during the creation of a project, the tendering, bidding and pre-contract phase of projects; and 15 post-contract ones arising during detailed design, construction, commissioning and transition to operations. All these causes of failure are related to shortcomings in management in one way or another and always involve human input. These shortcomings are mostly at senior level: the principle stakeholders, including the client, the Bid team, design consultants, project and construction managers and services contractors. The common element is invariably the human input and not the technical processes.

Generally though, failure is referred to in terms of cost over-runs and programme blow-outs for various reasons and these causes of failure are analysed and explained to clients, owners and shareholders ad nauseam.

But why do project failures keep happening? Why can't major international companies and government authorities develop risk management and bidding processes that prevent project failures, or at least establish realistic parameters that minimise the extent of failure if things start to go wrong?

If control and reporting is efficient and up to scratch all the time then properly informed decisions and mitigation should substantially reduce the risk of major failure.

Global Construction Success, First Edition. Charles O'Neil.
© 2019 John Wiley & Sons Ltd. Published 2019 by John Wiley & Sons Ltd.

However, this only applies if the project has been commissioned on *'solid foundations'* and unfortunately many are not.

Projects that fail generally do so for a combination of reasons and in this chapter we analyse these different reasons in depth. However, the one common denominator amongst all these reasons is human error at some stage or other in the process, as shown in the following examples.

The global construction and civil engineering industry is notorious for its slim margins, typically in the 2–3% range, which are brought about by the competitive bidding process; this leaves no room for mismanagement, but fine margins are not a prime cause of failed projects.

So, our interest is in finding out why some projects are very successful and why others have catastrophic programme overruns and financial losses; losses that can vastly exceed the original contract price.

The '35 common causes' form the second part of this chapter, but first we will review some high profile projects from the last two decades that have run into serious trouble.

The following projects are large-scale examples in four countries, but there are many more well-known examples around the world of projects of all sizes. Also refer to Chapter 15 for similar examples of Public Private Partnerships (PPPs) that were poorly planned and vastly over budget and programme.

With the following projects we are talking about large cost and time overruns, not the marginal losses that often occur in projects but which do not put the existence of the head contractor at risk.

4.2   High Profile *'Problem Projects'* Since 2000

Politically motivated, poorly planned, vastly over budget, behind programme.

In this section we review and make brief comments on the key issues associated with the following international projects that ran into serious trouble. These issues show the commonality of problems that can occur with the implementation of major projects. All of these projects were government sponsored, and all of them were big enough to attract media attention and necessitate rigorous post-mortems, so the information we use is all publicly available on the internet.

- Berlin's New Brandenburg Airport
- The Central Artery/Tunnel Project (the Big Dig), Boston
- Brisbane Airport Link
- The Scottish Parliament Building
- Stuttgart 21 Rail Development
- National Physics Laboratory, UK
- Metronet, UK

   Followed by a special comment on two iconic buildings:

- The Sydney Opera House
- The Hamburg Elbphilharmonie.

With such large projects there are a number of risk areas that can potentially be a problem, such as with the planning, financial feasibility, cost estimating and with the construction process, etc. However, there are people driving all these areas, so it is more than likely that our old friend *'erroneous human error or behaviour'* is behind any serious problem with any of these components. The scope of responsibility can extend over a broad range of participants, from the boardroom and government departments; to design and environmental consultants; to cost planners and programmers; and at the coalface on the construction site. We will encounter all sorts of people from these areas as we progress through the following pages.

The key issues involved with these projects remain the same for all size projects and there are numerous smaller ones in every country that have incurred the same problems.

4.2.1   Berlin's New Brandenburg Airport

Comment: this project is an extraordinary example of incompetent planning and implementation, on an ongoing basis. Industry professionals cannot understand why the authorities have not been able to address the identified problems and expedite the completion. 'After almost 15 years of planning, construction began in 2006. Originally planned to open in October 2011, the airport has encountered a series of delays and cost overruns. These were due to poor construction planning, execution, management and corruption. Autumn 2020 became the new target for the official opening date as 2019 became too improbable. A new TÜV report published in November 2017 suggested that the opening could even be delayed until 2021'. https://en.wikipedia.org/wiki/Berlin_Brandenburg_Airport

With a projected delay in opening of nine years or more, the anticipated total cost in early 2018 was €10 billion compared to the original budget of €3 billion including construction re-work costs and compensation to services operators and retailers. The primary cause was originally the failure of the project management to comply with fire regulations, plus a large number of functional design deficiencies that were unearthed late in the day. As the years have passed it has become quite apparent that incompetent management of the project has been the principal problem.

4.2.2   The Central Artery/Tunnel Project (the Big Dig), Boston

Comment: the $12 billion cost increase and nine year program overrun was the result of very poor execution.

'Planning began in 1982; the construction work was carried out between 1991 and 2006; and the project concluded on December 31, 2007 when the partnership between the program manager and the Massachusetts Turnpike Authority ended. The Big Dig was the most expensive highway project in the US, and was plagued by cost overruns, delays, leaks, design flaws, charges of poor execution and use of substandard materials, criminal arrests and one death. The project was originally scheduled to be completed in 1998 at an estimated cost of $2.8 billion (in 1982 dollars, US$6.0 billion adjusted for inflation as of 2006).However, the project was completed only in December 2007, at a cost of over $14.6 billion ($8.08 billion in 1982 dollars, meaning a cost overrun of about 190%) as of 2006. The Boston Globe estimated that the project will ultimately cost $22 billion, including interest, and that it would not be paid off until 2038'. https://en.wikipedia.org/wiki/Big_Dig

4.2.3   Brisbane Airport Link

Comment: this is one of the worst examples of a PPP motorway that was ill-conceived on the basis of extremely over-optimistic traffic projections.

'The big toll road operator Transurban has bought Brisbane's troubled Airportlink roadway for $2 billion – almost 60% less than the original $4.8 billion construction cost. The 6.7 km AirportlinkM7 tunnel connecting Brisbane Airport with the CBD was completed in July 2012. Initial forecasts from the failed project operator – the BrisConnections consortium – that the Airportlink would attract 170 000 vehicles a day within months of opening proved wildly optimistic.

Fewer than 50 000 vehicles a day were using the tollway by the time it went into receivership in early 2013'. www.abc.net.au/news/2015-11-24/brisbanes-troubled-airportlink-sold-at-almost-60pc-loss/6968586

'February 20, 2013 – The Australian Shareholders Association (ASA) has described Brisbane's Airport Link as the "biggest construction project disaster" in recent history, local media reported on Wednesday. Queensland toll road operator BrisConnections was placed in administration on Tuesday seven months after opening, with debts totalling more than 3 billion AU dollars (3.1 billion US dollars) according to the Australian Broadcasting Corporation (ABC). The receivership came after traffic flows only reached less than half of what was projected'.

4.2.4   The Scottish Parliament Building

Comment: this project was a fiasco; a perfect example of political bungling
'From the outset, the building and its construction have been controversial. The choices of location, architect, design, and the construction company were all criticised by politicians, the media and the Scottish public. Scheduled to open in 2001, it did so in 2004, more than three years late with an estimated final cost of £414 million, many times higher than initial estimates of between £10 m and £40 m. A major public enquiry into the handling of the construction, chaired by the former Lord Advocate, Peter Fraser, was established in 2003. The inquiry concluded in September 2004 and criticised the management of the whole project from the realisation of cost increases down to the way in which major design changes were implemented'. https://en.wikipedia.org/wiki/Scottish_Parliament_Building
   Also refer to Chapter 13 for a summary and further perspective.

4.2.5   Stuttgart 21 Rail Development

Comment: as of early 2018 the expected completion date and final costs are anyone's guess, with the project history inspiring no confidence in its overall management.
'Stuttgart 21 is a railway and urban development project.... Its core is a renewed Stuttgart Hauptbahnhof, amongst some 57 km of new railways, including some 30 km of tunnels and 25 km of high-speed lines....

The project was officially announced in April 1994. Construction work began on 2 February 2010

On 19 July 2007 it was announced.... Identified funding sources... also made provision for possible increases over the €2.8 billion estimate of up to €1 billion.... As of March 2013, total costs are officially estimated at 6.5 billion euros.

Heated debate ensued on a broad range of issues, including the relative costs and benefits, geological and environmental concerns, as well as performance issues. On 1 October 2010, the biggest protest so far took place with an estimated 100 000 people taking part in the demonstration against the project.

As of 2017, the start of operation is expected in 2021' https://en.wikipedia.org/wiki/Stuttgart_21

4.2.6   National Physics Laboratory, UK

Comment –: the lessons learnt out of this failed project are very important, even though several years have passed. Essentially, the contractors signed up to deliver a specification that it was technically not possible to build. The heavy losses from this project, before the contract was terminated, and those sustained on the Cardiff Millennium Stadium, contributed to the 2001 disposal of John Laing's construction division to O'Rourke for £1.

'Firms ran up £100 m loss on PFI laboratory scheme'
Wednesday 10 May 2006 09.20 BST

Private companies and banks lost more than £100 m on a Whitehall Private Finance Initiative that went wrong, a National Audit Office report reveals today. Builders John Laing and the Serco Group had to abandon a £130 m project to build a new headquarters for the National Physical Laboratory – a world-class facility to measure time, length and mass – in Teddington, Surrey.

The Teddington scheme went wrong when the contractors failed to take account of specifications needed for scientists and engineers to carry out delicate measurement work. https://www.theguardian.com/business/2006/may/10/highereducation.businessof-research

The Termination of the PFI Contract for the National Physical Laboratory

> *'The Department for Trade and Industry handled the termination of the contract well. But different handling of the project by all parties at the early stages might well have avoided the problems which led to the termination'*
> *Sir John Bourn, the head of the National Audit Office, 10 May 2006'''*
> www.nao.org.uk/publications/0506/the_termination_of_the_pfi_con.aspx

4.2.7   Metronet, UK

Comment: Metronet went into administration in July 2007, predominantly because its poor corporate governance and leadership meant that it could not manage its shareholder-dominated supply chain.

'The Metronet PPP contracts to upgrade the Tube left the DfT without effective means of protecting the taxpayer. Metronet's failure led to a direct loss to the taxpayer of between £170 million and £410 million. The DfT's work with the Mayor of London, TfL and London Underground on a long term solution will need to improve governance and risk management in the new arrangements they are intending to put in place to protect the taxpayer.

The Comptroller and Auditor General, 5 June 2009'

Transport for London (TfL) guaranteed 95% of Metro net's borrowing …… …When Metronet failed, the DfT made a £1.7 billion payment to meet the guarantee so that the running of the Underground would not be compromised.

www.nao.org.uk/report/the-department-for-transport-the-failure-of-metronet

4.2.8   The Sydney Opera House and the Hamburg Symphony Concert Hall (Elbphilharmonie)

Comment: these two world renowned buildings deserve a mention because they have quite a lot in common, even though they opened 40 years apart in 1973 and 2017.

- They were politically motivated and suffered considerable political interference.
- Were difficult to design and construct.
- They both ran heavily over budget and programme
  - The Opera House opened ten years later than planned and cost $102 m against its original budget of $7 m. It was paid for by a special lottery, so the taxpayers had no complaint.
  - The Symphony Concert Hall opened eight years later than planned and cost €789 m against its original budget of €241 m.
    However, this has long since been forgotten with the Opera House and no doubt will be with the Symphony Concert Hall, because of their unique and outstanding features that make them iconic world buildings:
- The Opera House is architecturally amazing and in a superb location and this makes it a major tourist attraction even if you never go inside.
- The Symphony Concert Hall, nicknamed Elphi, is acclaimed as being one of the largest and most acoustically advanced concert halls in the world; all the 2500 seats are within 30 m of the orchestra. It is also situated in a superb location in Hamburg Harbour.

There has been much written about the substantial issues encountered during the creation of these two majestic buildings, including the resignation mid-stream of the Opera House's Danish architect Jørn Utzon after clashes with the State government:

https://en.wikipedia.org/wiki/Sydney_Opera_House
https://en.wikipedia.org/wiki/Elbe_Philharmonic_Hall
Were any lessons from Sydney applied in Hamburg?

The simple answer is that there would have been no reason for lessons from Sydney to have been applied in Hamburg. These two landmark buildings had their own individual issues with planning, design, construction, cost and programme over-runs, and management by government and city authorities. There is no reason why the Hamburg authorities would have reflected on the Sydney experience. Similar mistakes may have been made, but they had no connection.

4.3   The 35 Common Causes

Given the above outcomes you would think that far more research, thought and practical effort would be dedicated to implementing more effective commercial risk management systems in order to reduce the probability of failure. Invariably the fault does lie with senior management, be it corporate or government. One clarification that I would like to make is that the primary causes of project problems that we discuss are principally related to commercial and site risk management as distinct from health and safety risk, which is vitally important but rarely causes major project failure.

In all cases it is actually a failure in the project management of one stakeholder or another that is the cause of the problems. Risk management systems should be able to anticipate and provide early warning or expose the problems as the project progresses through planning and into construction, and rectify or mitigate them to an acceptable extent, but clearly this did not work effectively in the above projects.

With government-backed projects some commentators say that it is not surprising they get out of hand because politicians interfere too much for their own political purposes, don't understand the processes or the risks and don't let the professional public servants run the projects unhindered (shades of *Yes Minister!*)[1]

To an extent this is correct, but as a generalisation it is unfair, because there are many government projects around the world that have been successfully implemented by very professional government authorities, with complex projects completed on time, on budget and to a high standard of quality, with these projects covering civil engineering (roads and bridges, etc.) as well as general construction works. These projects include both traditional design and construction (D&C) and PPPs.

The most common criticism of government projects is that government authorities have a perennial habit of underestimating the total cost and the programme, with figures being bandied about of approximately 20–30% for each. Such underestimating is a sure sign of inexperience or incompetence, or even *'good'* politics – i.e. if you knew the *'real'* out-turn cost, then projects wouldn't get off the ground. There would not have been a Sydney Opera House!

'Optimism bias' in public sector projects is not uncommon and creates unrealistic expectations by under-estimating time and cost and over-estimating the benefits. Therefore, when governments are looking at taking on innovative projects or ones with a higher risk profile, they should ensure that their investment decisions are based on realistic estimates and assumptions and that they have identified the potential risks as far as possible, with appropriate cost contingencies and mitigation plans.

> *Watch out for 'confirmation bias'*
> *– the notion that expectation plays a large part in comprehension,*
> *which can lead to subconscious poor decisions*
> *(it is part of airline pilot training)*

Many government authorities rely heavily on industry professionals to advise them and sometimes this advice has been found wanting in major projects that have run into trouble. Then the blame game starts, with the politicians and bureaucrats saying they were badly advised and the industry professionals saying that their advice wasn't heeded.

While it is easy to criticise government sponsored projects for wasting taxpayer's money due to poor planning and execution, it should not be forgotten that there have been many major projects that have been fully within the control of the contractor, that

---

1 *'Yes Minister'* was a popular BBC comedy in the 1980s.

have been won on a lump sum Bid submitted after all the checks and balances have been applied at all corporate levels, and they still incur huge losses. There are too many examples that show how this can and does happen.

The following is the detailed list of the 35 common causes of project failure that we have identified.

The most crucial of these causes mostly occur during the initial planning stages when the project is being structured, the concept and feasibility, planning approvals, bidding, design and cost planning, construction, commissioning and transition to operations, and rarely after the facilities have commenced commercial operations.

We have only provided explanations and examples for the items that we consider most critical and generally the most common causes of failure. Examples of all the items listed would fill a book on their own, but they are not necessary because they are either well understood by readers involved in the industry or they are self-explanatory.

4.3.1   Structuring, Bidding and Pre-Contract Phase

1) Rushing into a project without it being properly structured, with inadequate planning and organisation from the start, and all too hurried, e.g. the tender going out too early, or the contract being awarded too early, means that the budgets and financing, design, construction and operational requirements may not have been properly defined. Common to this sort of approach is a lack of a structured management platform and effective controls being put in place. The inevitable result will be increased costs, high claims by the contractor and delays in construction.

2) Failing to run design development *'hand-in-glove'* with cost planning can lead to very costly errors in D&C projects. This can apply with the schematic design during a D&C Bid and also with the detailed design after signing the contract.

   This is an inherent and significant risk in the construction industry's fragmented and non-integrated way of working. On every project, the team faces the eternal question that keeps everyone awake at night, *'How do you know that the costing is keeping up to date with the progress of the design and that the design being put forward will be competitive in price, and that everything has been included in the costs and that the estimators know that they been told everything?'*

   It is fundamental that D&C design development and cost planning run in close parallel so that different design options can be priced immediately and that there is ample opportunity for value engineering, all in the context of the targeted total Bid price that the Bid director will be aiming at. With large projects and tight Bid programmes it is a real trap if the design team gets too far ahead of the cost planners, because if it is found that the total Bid price is looking too high and the submission deadline is rapidly closing then there is no time left for the design team to consider other options and remedy the situation. This is where tight Bid programme management becomes so important.

   After commencing the contract there is not as much urgency, but a lot of time can still be wasted if the detailed design development is not closely monitored for cost against the contract price. With PPPs, client user groups often see this phase of the design development as an opportunity to get more for their money and the design and cost managers have to be careful to stick to the specifications they offered in the Bid.

3) Failing to properly involve the future facility managers in the design process, both during bidding and in the detailed design development, applies in all design and construction projects where there is an obligation on the developer or contractor to provide the client with a facility in a fully operational condition and where the client will be contracting out the facilities management (FM) to a specialist FM company, or even when the client is planning to manage the FM themselves, in which case the client needs to participate in the development of the project as if they were an independent facilities manager. For the sake of the exercise, assume that the project will have a 30 year life cycle in the same manner as a PPP.

The most common cause of problems in this area is not involving the facilities manager early enough and fully enough. When this happens:

- Design development does not benefit from the expertise of the facilities manager in terms of materials and equipment selection, serviceability and optimum life cycle, e.g. hypothetically, 'granite can be cheaper than bricks in the long term' or in practical terms, a higher capital cost today for a more efficient system may prove to be the best investment in the long run when daily running costs and life cycle replacement are both considered in the financial modelling.
- The facilities manager will have to accept what they are handed, and if it is found down the track that some of the services are difficult and expensive to maintain then they are in no position to complain or try and recover unexpected costs.
- Without this full involvement it is almost certain that items and costs will be missed or there will be poor procurement selections, to the detriment of both the client/user and the facilities manager.
- Additionally, it is likely that the scope of the future services in the FM subcontract will be inadequate and lead to disputes down the track.

The facilities manager therefore needs to be fully involved from the Bid stage, right through D&C and in the completion, commissioning and transition when they take over responsibility.

This is such an important area of risk management that we list the following indicative areas in which the facilities manager should be involved.

Bid phase:
- Technical and operational evaluations
- Selection and pricing of building systems, M&E, IT and security systems
- Systems efficiency, maintenance and life cycle implications
- Undertake energy modelling

D &C phase:
- Participation in the design development and user group process
- Participate in the health, safety and environment policies
- Development of policies and operational manuals
- Develop systems for the control and reporting of the services
- Build the asset database for the facility
- Sign-off all design and procurement

Commissioning and transition to operations:
- Train and induct the FM and client staff
- Witness all commissioning and testing of building systems
- Create crisis communications and management plans

- Formulate drills for emergency evacuations
- Stock and equip consumable items and maintenance stores
- Ensure the asset register is accurate and up to date
- Arrange FM/client coordination and communications on a daily basis

The above lists are by no means comprehensive, but are intended to alert managers to the areas to be covered generally, and the potential risks that will arise if any key areas are overlooked.

4) <u>Not signing up subcontracts with firm prices before Bid submission or signing of the contract</u> is really gambling. A contractor can only minimise their cost risk if they have pricing commitments from the majority of their suppliers and subcontractors in accordance with the design and specifications, and are confident that their cost estimates for those not locked-in are conservative and will cover the situation. *Subcontractors are a wily bunch and they will quickly realise if a head contractor is exposed. It is a dead give-away if RFPs are issued after the main contract has been signed, but you will be surprised how many contractors still do it.* A risk management check in sufficient time before the bidding process is complete should pick this up.

5) Insufficient investigation of key risk areas pre-Bid or pre-contract. Ignore at your peril:
   - Fire regulations and compliance (make the Fire Regulatory Authority your best friend)
   - Health, safety and environmental provisions
   - Time and cost involved in commissioning, completion and QA sign-off.

   There are a number of areas where both employers and contractors have been badly caught out on a regular basis over the years.
   - Employers because they have not done sufficient investigation of potential risk areas or made adequate allowances in their budget before entering into a contract, or because they have not included adequate terms in the contract for passing on potential risks.
   - Contractors that don't have methodical processes can miss expensive items in the rush to submit the Bid, including discrepancies in the contract documents that end up going against them.

6) <u>Insufficient written qualifications in the Bid</u> concerning key risk areas in the contract documents, which include the drawings and specifications.

   More often than not bidding is a stressful process and many items are left to be finalised at the last moment, out of necessity. One of the risks arising from this for contractors is that qualifications to the Bid will be overlooked in the final rush to the line. This can cause disputes down the track when the contractor says *'but I didn't allow for this'* and the client responds *'sorry but you did not qualify that in your Bid'*. With large Bids it is important that there is a designated risk manager who takes responsibility for collating all the risk management items and checks with the team leaders of all the other segments of the Bid. *(Refer to Chapter 17, team structure organisation chart for risk management.)*

   Qualifications can also work against the client's interest and budget if they are included by the contractor, but the client doesn't readily pick them up.

A Simple Example from A Real Project
The contract was for the supply, fabrication and erection of structural steelwork for infrastructure at an Australian coalmine. The tender drawings were obviously incomplete, consisting of basic site layout plans showing the structures required with plan dimensions, schematic elevations showing dimensions and some drawings showing typical member sizes, spans and connections. The tender called on bidders to make allowance for the missing detail, of which there was quite a bit, and submit a lump sum price for the completed works, based on their knowledge of what should be required and answers to questions they were invited to submit.

The contractor submitted a non-compliant Bid, clearly stating in the covering letter that their price was for a given tonnage of steel erected, broken down into the categories shown on the enclosed schedule with matching rates, and that any tonnage supplied over and above would be charged as extra. The written qualifications were clear and unambiguous.

The final price increased by 20% over the Bid price and the client initially refused payment, but on legal advice paid in full prior to the court hearing.

Risk management lessons – the contractor protected their position well and the client's process was very slack in not picking up the qualified Bid.

7) <u>Unrealistic programmes offered in the Bid.</u> Inaccurate and unrealistic development plans and construction programmes create all sorts of repercussions for head contractors and their subcontractors, property developers, investment clients, lenders, and insurers. I won't detail these repercussions here because they are largely self-evident, and are dealt with in detail in other chapters of the book. There are several reasons why developers and contractors produce inaccurate and unrealistic programmes, ranging from inexperience, excessive optimism or incompetence, to more complex reasons that might have intended consequences, such as with the effect on cash flow.

It is such an important topic that we have devoted Chapter 24 entirely to it.

8) <u>Undue interference on pricing and/or programme</u> can come from internal or external sources, e.g. managers that love building legacy monuments, overbearing clients or shareholders.

People in authority interfere and over-ride risk management 'stop' signs for a variety of reasons:
- Long-term clients lean on contractors to reduce their price 'if you want to keep our business', rationalising that this is guaranteed work for the contractor.
- Large shareholders think the stock price will be boosted if the company announces a lot of new business, so they have a chat with their representatives on the Board, who have a chat with senior management.
- The company is having cash flow problems so senior management 'buy' some projects to try and stay alive.
- Senior management and the directors think this particular project will be great for the company image and/or the stock price: 'don't worry too much if it is not profitable because the rest of the business is going well'.

In these situations two things invariably happen:

- Middle management, who know the figures, are not informed of the real reasons and are left shaking their heads – and feeling insecure.
- Things don't work out for the best and the company incurs far greater losses than ever envisaged.

Sometimes, but not often, senior management do have some cards up their sleeve that will make the project profitable, but if so they should take the operational team into their confidence as soon as possible, e.g. on almost all projects the programme has a direct bearing on profitability and if it can be substantially shortened for some reason then the project might be financially successful, but it can also work the other way.

Risk management check points are put in Bid programmes for prudent reasons, so ignore them and over-ride them at your peril. More than one small or medium building company has been brought down by just one or two projects.

9) <u>Bidding with annual revenue being the primary driver instead of profits.</u> Many companies, large and small, believe that the higher the turnover they generate each year the greater the chance of making a profit because of the financial gains they will make through creating improved efficiencies in delivery after winning projects, claims for variations and the squeezing of consultants and subcontractors during the life of the projects.

With this in mind they cut margins, preliminaries and contingency allowances to the bone and chase turnover (revenue) as their primary objective, bidding and winning projects on little or no profit margin. They also justify this approach by convincing themselves that they will at least get some overheads for the head office even if the project only breaks even.

Now this may work sometimes, but the construction industry is notorious for how many projects end up being unprofitable even when healthy allowances and margins are built into the winning price. So if you are chasing revenue and winning projects on an effective nil-margin basis then you are ultimately playing with fire. If anything goes wrong with time, cost and quality control then those slim margins or a break-even price will rapidly turn into a loss that will not be covered by variation claims and squeezing the subcontractors. Generally, when this happens it is really bad news for the subcontractors.

On top of this the company balance sheet has to accommodate performance bonds, bank guarantees and retentions on the progress claims, so it is likely that the project cash flow will be negative overall if there is the slightest hiccup. If this situation occurs for a company with several projects running concurrently then it will only be a matter of time until their bankers run out of patience.

Chasing revenue without profit is therefore a dangerous game for a construction company that does it often.

It is interesting to note how often construction companies make press releases announcing a new project they have won and its value. It would be of far greater interest to shareholders if they also announced their projected bottom line for that project as well, albeit with some cautionary qualifications. This is a good form of risk management in itself.

10) If 'super profit' predictions over-ride established risk management procedures then this due to one of two types of managers: those that can only see 'blue sky', or those who are desperately trying to prop up the company when it is doing badly.

11) Inadequate cost escalation terms in contracts. *'Rise and fall'* is an intriguing contractual term and it is easy to think that *'fall'* never happens, but it does occasionally and can cause some losses against budget. However it is far more likely that inflationary increases in labour and materials *('rise')* will be harmful and this can cause serious financial loss if not properly covered by the terms of contract. There are various formulae for this such as using a consumer price index.

12) Inadequate currency change terms in contracts. It is conventional wisdom that one of the quickest ways to go broke is to speculate on currency markets. With contracts involving cross-border currency transactions it is equally risky to have unhedged exposure, so it is not only prudent but also essential to safeguard against currency loss through hedging.

13) Failure to benchmark total Bid value on a 'common sense' or 'logical' basis against comparable scope contracts.

How did this 'project' escape through all the risk management check points?
The project was a European motorway where the winning Bid was €400 m and the end cost was more than €600 m. The contract called for 38 km of 4 lanes plus 2 emergency lanes through mountainous terrain, with 5.7 km of tunnels (7 off), 5 large bridges from 130 to 400 m in length, and a large amount of rock blasting out of the mountainside; 80 structures in all including entries and exits.

Summary – on global comparison the €400 m Bid price would have been competitive and probably tight for relatively easy terrain with no tunnelling, extensive rock blasting and large bridges. In hindsight, it is inexplicable how a quick and easy benchmarking like this did not stop the Bid consortium in their tracks. The end cost was more than €600 m and did not contain any margin.

Responsibility for the bidding and signing-off this project lay at several levels, from the estimators through to senior management, any of whom could have applied simple benchmarking. Is it possible that some more junior parties did highlight the problem but the *'grown-ups'* weren't listening?

---

14) Underestimating site and head office administrative costs and the time specifically required for the project, i.e. poor budgeting. This is a common error, as people tend to be over-optimistic about the resources and time required to run a project, especially during the commissioning and completion period. Variations and extensions of time with daily costs can cover their own administrative costs, but they will only partly alleviate basic miscalculations of programme and resources for the overall project. There is no substitute for experience in this area of budgeting. PPPs in particular are often underestimated due to the demanding requirements for contractual delivery.

15) Incorrect taxation assessments. Tax is a field for the experts, but even they make mistakes when complex financial structures are involved, such as with PPPs. Tax liabilities and tax credits can both be miscalculated or overlooked and an independent second opinion is recommended as a prudent matter of course where large sums are involved.

In one large PPP investment a new finance director joined the concession company and six weeks after he arrived he advised the board that he had found a $15 m tax credit that could be claimed from three years previously. This had been missed by internal accounting, the external financial advisors and the auditors. Now that was worth an early bonus!

16) 'Clever clients' who think they can do it better themselves by managing direct subcontractors instead of wrapping the project into a lump sum contract with the primary risk being taken by the head contractor.

The Squaire Building, Frankfurt Airport, Germany

Comment: it is a recipe for disaster when 'clever clients' think they can do it better themselves by managing subcontractors directly instead of wrapping the project into a lump sum contract with the primary risk and interface issues being taken on by the head contractor. The final cost of the project was reported as approximately EUR 1.3 billion, double the original estimate of €660 m. As a property investment, this was a disastrous result.

'The Squaire is an office building in Frankfurt, Germany, built between 2006 and 2011 on top of Frankfurt Airport long-distance rail station. The building is 660 m long, 65 m wide, 45 m high, and has nine floors. With a total floor area of 140 000 m^2 (1 506 900 sq. ft.) it is one of the largest office buildings in Germany. Its dimensions and design make it a 'groundscraper''. http://en.wikipedia.org/wiki/The_Squaire

---

It is a problem when *'you don't know what you don't know'*

---

17) New techniques or technologies. Using high risk new techniques or technologies that have not had sufficient proving-out can cause big losses after winning the contract.

Projects are launched fairly regularly that have extremely high technical demands that are *'ground breaking'* and when successfully completed will also have a high profile. If the specifications for which the contractor has to guarantee performance delivery require new and unproven technology then all the danger signals should be flashing.

However, the enthusiasm of senior management to win such projects and gain the high profile kudos on completion can sometimes overshadow and outweigh their pre-Bid investigations into how advanced and proven this new technology is.

This is not always a simple situation and depends heavily on the technical people being entirely open and frank about the status of the technological development, because it is likely that the contractor's senior management will not have an in-depth understanding of the new technology. The technical people might be over-optimistic because this is their *'big opportunity'*, or they may have only proven the technology in laboratory testing and be taking a leap of faith that it will work in a commercial situation.

The obvious answer is that the risks are carried by those directly responsible for the technical delivery and performance, and the contracts and subcontracts should be drawn up carefully to reflect this, with appropriate guarantees and indemnities.

Clients will take the view that *'if you promise to deliver the contract and you do, you will get paid, but if you don't deliver then it is going to cost you'*.

There is a wide range of industries in which this situation can apply in today's world of fast moving technical development, from prison security systems to *'clean room'* chip manufacturing factories.

Two highly expensive and publicly known examples in recent years have been the Siemens Particle Therapy Oncology Hospital, Kiel, Germany (http://www .marketwatch.com/story/siemens-ends-particle-therapy-project-in-kiel-2011-09-14; http://www.siemens.com/press/en/pressrelease/?press=/en/pressrelease/ 2011/corporate_communication/axx20110983.htm) and the National Physics Laboratory, UK – refer to the start of this chapter.

Sometimes client specifications and expectations are unrealistic and sometimes suppliers and contractors offer more than they prudently should.

18) Failure to *'move with the times'*. Conversely, the failure of project management or technical people to *'move with the times'* and use more efficient technical or management procedures that have been proven can lose you Bids.

19) Poor communications and relationship management with client, government authorities, subcontractors and suppliers can get you off on the wrong foot and it is hard to recover. **Listen hard to what the client is saying** – it can win you the project. The same applies with the community and media after winning the project.

20) Political risk. This can be difficult to assess, but many major contractors have incurred heavy losses when governments pull the rug towards the close of bidding after a great deal of time and cost has been expended by the bidders. This is a breach of good faith. Bidders know that only one of them will win and that is fair competition, but they certainly do not appreciate being strung along. It helps if the particular government agency agrees to pay compensation, but it does not happen very often and then it is mostly only a nominal amount.

4.3.2 Design, Construction, Commissioning, and Transition to Operations

The remaining causes of project failure are in this stage.

21) Failure to line up a competent project team sufficiently ahead of the project award, especially the project director (this applies equally to D&C and PPP teams).

22) Lack of continuity of personnel from Bid to D&C – and through to operations if it is a PPP. When this happens it is all too easy to blame others for problems arising.

23) Hiring inexperienced and under-qualified management resources is a serious trap and false economy.

24) Too many contract managers hired in, because many have no real company allegiance or loyalty and it is just another job. Even worse, if they get a better offer mid-way through the project they are likely to leave at short notice.

25) Project directors and construction managers in key positions must be good communicators. Poor communicators cause all sorts of problems. There are several references to communication throughout this book.

26) Denial, stubbornness and bloody-mindedness by headstrong *'sledge hammer'* construction managers can be really damaging if they do not identify and report potential risks at the earliest possible time, but keep the problem to themselves in the belief that they can fix it before it gets worse or before anyone above finds out. Managers with this sort of attitude cause real relationship damage in any event and should not be tolerated.

The construction industry unfortunately has an inherent culture of breeding a proportion of dominating 'sledge hammer' site managers who are also nearly always bad communicators, and yell and scream at everyone. This type of manager invariably ends up with poor relationships with clients and subcontractors, more than a reasonable number of disputes, poor team spirit and a lack of respect from their own staff.

Compare this to the real leader who communicates well with everyone, mentors and motivates, has all the skills to show the way, commands respect from all stakeholders, and does not need to raise their voice.

27) <u>Slow start</u> (particularly if the program is tight anyway).

28) <u>Fast start</u> – fast-tracking with inadequate preparation, resulting in unbudgeted costs and most probably costly rework. (See Chapter 24.)

29) <u>Basic changes to design after commencement</u> – this is costly for both the client and the contractor; the client has to pay for the design changes and may well have a delayed completion and commencement of revenue, and is costly for the contractor because of the delays incurred, the true costs of which are rarely fully recouped.

30) <u>Failing to accurately track and document design and equipment changes</u> and getting them signed off, and also failing to raise timely cost variations.

31) <u>Insufficient independent in-house or external overview (see Chapter 27).</u>

32) <u>Poor subcontract management and payment</u> – can cause serious construction delay (see item 35 below).

33) <u>Undercapitalised subcontractors</u> who cannot survive slow cash flow or a significant loss on a particular project.

34) <u>Head contractors that have inadequate financial capacity</u> – contractors need to have sufficient working capital for the size and complexity of the project. Major investors and governments can get caught out by contracting with contractors, particularly tier-two contractors, who might initially appear to have the financial capacity, but then find that the size and complexity is greater than they can manage and finance, or the head contractors take on a multiplicity of such contracts to the extent that they lose control of their business, as with Carillion in the UK in 2018.

35) <u>Clients and head contractors that are chronically bad payers</u>. It is not uncommon to come across clients that simply refuse on principle, and without any proper contractual justification, to make full and timely payments even though the progress and variation claims have been independently certified for payment. This type of client likes to find spurious reasons to only pay say 85% of each certificate and then mostly later than the date stipulated in the contract. The reasons for such obstinate and frustrating behaviour can be many and varied, and often as not it is cash flow driven but not always.

Some head contractors do the same with their subcontractors, but they overlook the fact that subcontractors and suppliers generally respond in direct proportion to the way they are treated and if it is really unreasonable then the bad payer will pay more in the long run, one way or another. This sort of practice can be really damaging to relationships, but conversely if payments are made promptly then the subcontractors will often go the extra mile for their client and work harder to meet the programme, and possibly not bother to claim for some unexpected costs they have encountered. It is a two way street and it takes only a small amount of goodwill to generate significant performance improvements.

Contractors do not have a problem with claims that are challenged for valid reasons provided there is an open and honest debate about the situation, but clients that short-change and are late with every payment as a matter of course are infuriating and are of course in breach of good faith as well as breach of contract.

Of course contractors are not angels either and often put in grossly inflated claims or claims for work that is not completed or where the milestone has not been reached.

4.4 Project Leadership – How Bad Can It Get?

Lessons to learn from grossly incompetent site management

Note: for legal reasons the identity of the following project cannot be disclosed, but the facts described are an accurate description of events.

The following story describes a project that ran off the rails well and truly and how it was brought back on track. It all went wrong because the project director was completely out of his depth and also had the wrong personal skills to be leading the project. It demonstrates the need for robust risk management processes at all stages in a contract to protect the client, the contractor, subcontractors, and suppliers. **This factual account shows how project leadership is vital to success.**

This was an industrial D&C with a contract value of $140 million and a 3 1/2 year programme. It was not a large project in itself but a vitally important one because it was a critical component in a much larger complex, with the rest of the industrial complex having activities that fed off this facility. The scope of works was very detailed in respect of the deliverables and the contract was tight, especially in regard to the programme, with heavy liquidated damages to a cap of 40% of the contract value, recoverable after five months of delay subject to other extenuating circumstances.

The contractor was a subsidiary of an international company who were recognised leaders in this field. They had no other involvement with the rest of the industrial complex, but were engaged because of the criticality of this particular facility. The contractor hired an outside construction director rather than appointing someone from inside the company and they commenced on site with a suitably experienced team of 14 personnel.

Three months before the due completion date the client raised urgent concerns about the progress whereupon the contractor advised they would be four months late and subsequently produced an updated programme. Two months before the new completion date the client concluded that this date was clearly not possible. The contractor then advised that they would require another seven months, i.e. completion would be 11 months later than the original contract completion date.

The client demanded urgent expediting action, so the parent company of the contractor appointed an external audit team to review the entire project and make recommendations. For the parent company, this contract value was actually quite small in the context of their annual global revenue, so they had not been keeping a close eye on the project. For the client, the liquidated damages that would likely apply were far less important than the serious flow-on effect to the overall industrial complex.

The independent audit team appointed consisted of a project manager, quantity surveyor and planner. Whilst all were very experienced and had been involved in many difficult projects internationally, they were taken aback by the problems on this particular project.

- They were introduced to the construction director by the national managing director, but immediately after he left the site the construction director said he would throw them off the site if they did anything without his approval. After a swift tour of the site he then declined any sort of cooperation – no programmes, no subcontractor information, and no access to interview team members on site. So, the managing director was quickly recalled by the audit team!
- Behaviour only marginally improved. Senior members of the site team were threatened with dismissal if they talked to the audit team, so the managing director was recalled again!
- The audit team spent four days reviewing the status of the works: the subcontractor arrangements, cost reports and interviewing the key site management personnel, following which they produced their initial findings and recommendations, including:
  - The master programme dated five months earlier had not been revised since and was showing a number of activities as completed that had not even started. The full-time planner/programmer said he was only allowed to produce what the construction director instructed – and he was not allowed to check progress on site. The construction director had for some time been seriously misleading the client and his own head office about the progress by delivering false programmes.
  - There was a similar situation with the project accounts, which showed many dubious counter-claims against subcontractors as being fact and, with these assumptions, indicated that the total project costs were within the original budget. This was clearly misleading and the end result was a 50% cost over-run, not including substantial liquidated damages.
  - There were 22 specialist subcontractors and, properly resourced, there should have been 700 workers on site, but 18 of the subcontractors had no personnel on site and the remaining 4 had about 30 between them.
  - All of the subcontractors were in dispute with the contractor for a wide variety of reasons, including unreasonable back-charges. The construction director had stopped payment of all claims in an attempt to force the subcontractors to concede on the disputes, with these stopped payments including a substantial number of standard progress payments certified by the engineer.
  - The anticipated final cost report failed to reflect the extent of costs and liabilities to complete the works.
  - All but four members of the site team were working on the disputed claims and back-charges on the instructions of the construction director. In many cases spending a lot of time trying to build a case when there wasn't one.
  - The four site managers who were actively trying to keep the site going were capable people, but they were all on the point of walking out through sheer frustration.
  - Several of the activities that had not been started could not start because they required planning approval, with some needing a three to four month lead time, and the applications had not even been prepared, let alone submitted.

The construction director was obsessed with 'squashing' subcontractors. He had completely lost sight of delivering the project. He admitted in interview that he had only read the scope of works and had not read the contract thoroughly at any time in the past three years, and stated that it was unimportant because he knew what was required technically. Clearly he had no idea of the obligations under the contract and did not seem to comprehend the significance to the overall complex.

It turned out that whilst the construction director was an experienced and capable design engineer he had never actually undertaken project management and had no idea about efficient methods and processes.

He camouflaged his lack of experience by heavily bullying his team, the suppliers and subcontractors, and would not tolerate any other opinions, constructive or dissenting. His arrogance and rudeness had to be seen to be believed.

He was relieved of his duties a few days after the initial report from the audit team and replaced by a properly experienced construction manager.

The independent audit team rapidly made several other expediting recommendations, with the priorities being:

- Restoration of the subcontractors' cash flow and freezing of the disputed claims and back-charges, pending a realistic review of them. The restored trust of the subcontractors took high importance.
- Urgent attention to planning applications and approvals.
- Acceleration of works by all the subcontractors on a 24/7 basis (the cost was actually far less than the potential damages).
- Creating realistic and accurate programmes that were effective as management tools, both the master programme and separate component programmes.
- The project achieved commissioning and practical completion four months after the audit team was engaged to the extent of effective integration with the larger complex, although some minor less important works still took a few more months to close out. Needless to say, the contractor suffered heavy financial losses and some loss of reputation, although the latter was somewhat recovered through taking the drastic actions required to recover the situation.

4.5 Lessons Learnt from Incompetent Site Management

- The leadership skills of the construction director are paramount and must cover knowledge of the contract, technical and programming competence, methods and processes, as well as the important personal skills of communications and relationship management that engender motivation and team spirit.
- If efficient risk management processes had been in place the alarm bells would have started ringing a year or more earlier and the parent company would have had the opportunity to rectify the situation and instigate changes to enable completion on time and to mitigate the potential losses.
- A project may seem small in the overall context for an international company, but even 'small' projects have the potential to cause heavy financial and reputational loss.
- The client's project monitoring had also been quite slack, although they did identify the problem and demand action from the contractor's parent company. Clients can protect themselves inexpensively by engaging independent project auditors to monitor on their behalf.

4.6 Conclusion

When you really examine the common causes of project failure you might conclude that it really should not be too hard to reduce or eliminate these causes and that all we have to do is pass on the lessons learnt to the next generation of managers.

However, this has not been the case to a sufficient degree. With most of these common causes the underlying problem is the human input at either the corporate or the project level, or both, and some sectors of the industry continue to have difficulty in coming to grips with this fact of life.

5

The Use and Abuse of Construction Supply Chains

Professor Rudi Klein LLB, FCIOB, FRSA, Barrister, Cert. Ed

CEO at Specialist Engineering Contractor's (SEC) Group, UK

> *Money is like muck, not good except it be spread.*
>
> Francis Bacon, English lawyer, courtier and philosopher in *Essays* (1625)

5.1   Introduction

The UK construction industry has long had a worldwide reputation for its prowess in design, engineering, and project management skills. However, its ability to deliver high quality infrastructure and construction projects within cost and on time is severely hampered by embedded and outdated practices that add cost rather than value. The delivery process is often characterised by fragmented and disconnected inputs that detract from prioritising the needs of the project and the requirements of the client or customer.

There have been two recent events in the UK that have highlighted some of the dysfunctionalities in the construction delivery process. The first was the closure of 17 Edinburgh schools in 2016. This followed the collapse of nine tonnes of masonry at a primary school early in 2016. It was extremely fortunate that there were no injuries or fatalities amongst the children. The independent inquiry into the collapse confirmed that it was the result of poor quality construction. This was not solely the fault of one contractor but was the result of a poorly managed delivery process driven by a lowest price culture. Professor John Cole, the author of the *Report of the Independent Inquiry into the Construction of Edinburgh Schools* (published in February 2007) recommended – amongst other things – that best practice methodologies be introduced into the procurement process to optimise design and construction quality.

Then came the Grenfell Tower tragedy. On 14 June 2017 a fire engulfed a 24-storey residential tower block in West London; 72 residents were killed. An independent review of the regulatory framework around building safety was set up.

It was led by Dame Judith Hackitt. In her final report she had this to say:

> Payment terms within contracts (for example, retentions) can drive poor behaviours, by putting financial strain into the supply chain. For example, non-payment of invoices and consequent cash flow issues can cause

Global Construction Success, First Edition. Charles O'Neil.
© 2019 John Wiley & Sons Ltd. Published 2019 by John Wiley & Sons Ltd.

subcontractors to substitute materials purely on price rather than value for money or suitability for purpose.

Para. 9.11, *Building a Safer Future – Independent Review of Building Regulations and Fire Safety.*

These events, together with the collapse of global contractor Carillion at the beginning of 2018, shattered the image of UK construction. Various inquiries into the collapse of Carillion revealed its contempt for its supply chain and the extent to which it abused those firms carrying out its work.

Some would argue that construction is only as good as its clients. There is, no doubt, some truth in this. Many client organisations – in both private and public sectors – are ill-advised by their consultants and legal advisers who tend to go down the path of least resistance by sticking with what they know. As a result the emphasis continues to be on driving down price and ensuring that all risk is laid off to the supply chain.

5.2 Construction: An Outsourced Industry

There is no other industry sector that outsources product/or service delivery to the extent seen in construction. According to research carried out for the UK Government Department for Business in October 2013 the average project in the UK deploys between 50 and 70 tier-two contractors (subcontractors). The contract packages can be as small as £20 000 but most will average £50 000. There is further subcontracting to levels three, four, and even more. The delivery process has become characterised by disintegrated, hierarchical, and sequential inputs.

As a consequence the cost of UK construction exceeds that in most of the European Union member states. Every project participant will seek to recover its costs and margins together with contingencies to take account of possible risks. Each will have taken out project-related insurance policies such as professional indemnity and contractors' all risk policies, which all cover the same type of risks. Furthermore all the myriad interfaces have to be coordinated and managed.

In this scenario the incidence of disputes is high. The lack of an overall and robust risk management strategy, which ensures that the design and construction will deliver the value demanded by the client, inevitably means that some risks will fall into 'black holes' or are simply ignored in the hope they will eventually disappear. Much time, effort and cost is then expended in seeking to allocate responsibility when the risks in question actually materialise.

All this is compounded by the fact that the largest UK contractors are severely under-capitalised. Under one legal test of insolvency they could be described as insolvent if the value of their assets are exceeded by that of their liabilities. For many years their business models have been shaped by reliance on their ability to drive down the prices of firms in their supply chains and to manipulate their supply chains' cash flow. Generally they are not in a position to meet the claims from their supply chains before they have received the necessary funding from their clients.

5.3 Adverse Economic Forces Bearing Down on the Supply Chain

There is a double economic 'whammy' for construction supply chains. First, they may have to subsidise the cost of construction because lead contractors will often not be in a position to pass on to their clients the full costs associated with the extent of the outsourcing on particular projects. Second, they will often have to subsidise the lead contractor by extending trade credit and, thus, ultimately, funding the project. According to research carried out for the UK Business Department referred to earlier, tier-one contractors were found to be net receivers of trade credit whereas tier-two firms were net providers of trade credit.

> The undercapitalisation of tier-one contractors and their reliance on business models generating negative need for working capital may in practice be as decisive as industry attitudes to risk and business models of risk transfer in determining whether intended changes to procurement and form of payment (milestone payments, performance-based contracting) will be taken up by the industry, and with what effect on competition.
> (*UK Construction: An Economic Analysis of the Sector, Department for Business, Innovation & Skills*, July 2013).

Large tier-one contractors will also deploy other means to extract funding from their supply chains. These will include so-called main contractor discounts, rebates, pre-bates and profits gained through the application of supply chain finance arrangements (whereby firms in the supply chain have to pay a fee for earlier payment).

5.4 Supply Chain Dysfunctionality

Construction supply chains are not supply chains at all. Rapid mobilisation (and then dispersal at the end of the project) of numerous subcontractors at different levels of subcontracting – usually following selection on lowest price – can hardly be described as supply chains.

> The UK construction industry…is thought to be more fragmented than its major competitors such as Germany or France.
> (*UK Construction: An Economic Analysis of the Sector, Department for Business, Innovation and Skills*, July 2013.)

What are supply chains? Chains have links and therefore a fully functioning supply chain should, at least, have the following links:

- A communications link
- A cash link
- A people link
- A product link.

The communications link is about having regular and positive dialogue between all parties to facilitate efficient delivery of the project; not, for example, issuing early warning notices that state there is a risk of delay and the recipient is responsible.

Supply chains do not exist for the purpose of funding projects. Recently the Specialist Engineering Contactors' Group analysed the accounts of the top dozen UK construction companies. They were owed £1 billion of cash retentions by their clients but 85% of this was funded as an unsecured loan by their supply chains.

The people link is about genuine collaboration between different supply chain entities to realise the objectives of the project. The product link enables all supply chain entities to fully understand the clients' success factors with a shared commitment to ensure that those factors are always prioritised in the delivery process.

In other words a supply chain is a dynamic resource that should be deployed to minimise project risk from the outset of the procurement process, not a receptacle for dumping all commercial and project-related risk after the works have commenced. How many examples are there of projects having a risk register that has been signed off by key elements of the supply chain before commencement of construction? Probably very few.

In fact this picture has not significantly changed since the 1970s when architect Professor Denys Hinton described the construction in the following vivid terms:

> …the so called building team. As teams go it really is rather peculiar, not at all like a cricket eleven, more like a scratch bunch consisting of one batsman, one goalkeeper, a pole-vaulter and a polo player. Normally brought together for a single enterprise, each member has different objectives, training and techniques and different rules. The relationship is unstable, even unreliable, with very little functional cohesion and no loyalty to a common end beyond that of coming through unscathed.

5.5   Addressing the Issues and Solutions

The UK construction industry has never been short of reports and initiatives that have fully described the problems and offered solutions. Examples from recent memory include Latham (*Constructing the Team,* 1994), Egan (*Re-thinking Construction,* 1998) and Farmer (*Modernise or Die,* 2016).

5.5.1   Latham

The late Sir Michael Latham's legacy was Part 2 of the Housing Grants, Construction and Regeneration Act 1996, commonly referred to as the Construction Act. The Act made a significant difference in improving payment security especially for small and medium-size enterprises (SMEs) in the supply chain. Previously, payment disputes could only be resolved in arbitration or by the courts. This meant that most SMEs would have had to sacrifice their entitlements rather than having to incur substantial legal costs in issuing arbitral or court proceedings.

The Act introduced a statutory right to refer disputes to adjudication. This was to be an inexpensive and quick procedure aimed at providing a stop-gap decision that would help the cash to flow. As the late Lord Ackner said in the debates on the Act in the House of Lords: '[Adjudication] was a highly satisfactory process. It came under the rubric of "pay now argue later" and was a sensible way of dealing expeditiously and relatively inexpensively with disputes . . . .'

Now contrast this with the comments made by Lord Fraser in the recent case of **Ground Developments Ltd v FCC Construccion SA & Ors** (2016) EWHC 1946. First the facts in the case:

Ground Developments Ltd (GD) had been engaged as a subcontractor by a joint venture comprising FCC Construccion, Samsung, Kier Infrastructure and Merseylink Civil Contractors (the joint venture (JV)) to execute certain groundworks in connection with the new six lane toll bridge over the River Mersey.

GD had submitted three payment applications totalling almost £200 000 plus VAT. The JV chose to ignore them; they didn't issue any notices. GD sought adjudication. Unsurprisingly the adjudicator ordered the JV to pay up. He also ordered the JV to pay all GD's fees (no doubt reflecting the fact that there was no meritorious reason for the JV denying payment).

GD was forced to go to court to enforce the adjudicator's decision; the JV defended the enforcement proceedings.

Their defence had nothing to do with whether or not the £200 000 was outstanding. During the adjudication the JV had maintained that the adjudicator had no jurisdiction because there was no contract when, in fact, they had ignored numerous attempts by GD to complete a NEC3 subcontract. During the enforcement proceedings – when pressed by the court – they changed their tune. There was a contract albeit on their terms. Mr Justice Fraser enforced the adjudicator's decision.

> Finally, excluding VAT, the parties collectively spent a total sum by way of costs in these proceedings in excess of £55 000, arguing about the enforceability of a sum of about £207 000, which was potentially repayable in any event because of the temporarily binding nature of adjudication. That is over one quarter of the sum the subject of the decision. It cannot pass without comment that this is contrary to the purpose of Parliament when it imposed this alternative, and temporary, process of dispute resolution upon parties who enter into construction contracts.

The £55 000 would not have included the amount spent by the parties on legal fees incurred during the adjudication.

Therefore a key element in the statutory framework initiated by Latham has lost some of its original gloss. Adjudication is essential for dealing with the everyday type of disputes that revolve around set-offs, often spurious counterclaims and recovery of outstanding retentions. However, its benefits have to some extent been subverted by the increasing costs relating to the fees and expenses of adjudicators and the legal costs associated with legal representation.

This is not to ignore some of the enduring benefits of the Act such as the limited ban on pay when paid, the right of suspension for late payment, and the statutory payment notice procedure introduced in the 2011 amendments to the Act.

5.5.2 Egan

Concerned about the inefficiencies and waste in an industry responsible for 7% of GDP the (then) Labour Government commissioned Sir John Egan (ex-chairman of Jaguar cars) in 1997 to report on how clients and the industry should change delivery models to ensure cost effective and value for money outcomes. Egan didn't pull his punches. Delivery had to be more integrated and collaborative. A follow-up report in 2001, *Accelerating Change*, sought to put in place a targeted programme of activities aimed at achieving an integrated delivery process. *Accelerating Change* was the product of a newly established Strategic Forum for Construction chaired by Egan and comprising representatives from government and industry:

> Payment practices should be reformed to facilitate and enhance collaborative working.

The forum was charged with producing models for payment mechanisms and key performance indicators for payment within supply chains to establish and benchmark best practice. Needless to say these initiatives either lost their impetus and/or were never implemented. This is the way of many good initiatives in the industry aimed at inducing long-lasting change. Both parties – government and even some industry bodies – have been guilty of blocking positive change where it doesn't suit their own vested agendas or, like a child with a new toy, have lost interest and moved on to something more interesting.

5.5.3 Farmer

In 2016 Mark Farmer was commissioned by the Construction Leadership Council (in reality the Business Department) to carry out a review of the UK construction labour model. The outcome, a report titled *Modernise or Die*, echoed the industry's failings that were canvassed in previous reports. Farmer put his finger on the reason for the lack of progress in improving the industry's delivery processes:

> There is no strategic incentive or implementation framework in place to…initiate large scale transformational change.

He advocated the use of project bank accounts (PBAs) and *'a move away from a culture of using other people's money to make money'*. The Government's response to Farmer was fairly non-interventionist. The industry had to sort its own problems out. Such exhortation is 'pie in the sky'. The industry is incapable of sorting out its problems unless there is in place some external driver.

5.5.4 The Plunder Goes On

In the meantime the plundering of supply chain financial resources, especially by the largest UK contractors, continues. All sorts of scams and whizzes are deployed to extract cash from supply chains or to create reasons for justifying late or non-payment.

So-called main contractor discounts are common but other impositions such as pre-bates or rebates (not necessarily geared to the volume of work generated) are slapped on unwilling or reluctant firms. Whether all this discounting is revealed to clients – especially where there are target cost arrangements in place – is open to question.

   In some cases conduct verges on sheer dishonesty. For example, subcontractors some-times find that they have been sucked into contracts under which they are being paid the correct amounts on time. This continues for a period. The firms are given more work. Then as time wears on – and after having been given a large amount of work by the same company – payments start to slow down. They become irregular and never reflect the correct amounts. Eventually the firm in question has to go into insolvency. The company responsible for this act of dishonesty then hoovers up the resources of the company to complete the work and gets away with not paying the vast amounts of cash that were owing.

5.5.5   Conditions Precedent

The *'condition precedent'* clause has now become *de rigueur* in many subcontracts. Often they are used to deny any entitlement to payment unless certain bonds or warranties are signed. However, contracts are entered into without subcontractors being given sight of the relevant documents that they may be required to sign up to. The example below is of a condition precedent clause, which is even more perfidious:

> Notwithstanding any other provision of this subcontract, no payment is due or is made…if performance of the obligations of the Subcontractor in terms of this subcontract is not in compliance with this subcontract or otherwise to the satis-faction of the Contractor acting reasonably.

   This, potentially, could provide the contractor with countless reasons for not making payment; even where such reasons lack substance.

5.5.6   Cross-Contract Set-Off

Cross-contract set-off is fairly popular.

> The Contractor shall be entitled to set-off against payments due to the Sub-Contractor any amounts which the Contractor has claimed against the Sub-Contractor or to which the contractor has become entitled to under any other contract between the Contractor and Sub-Contractor.

Section 110(1A) in the Construction Act outlaws such clauses:

> The requirement…to provide an adequate mechanism for determining what pay-ment become due under the contract, or when, is not satisfied where a construc-tion contract makes payment conditional on –
>
> a) the performance of obligations under another contract, or
> b) a decision by any person as to whether obligations under another contract have been performed.

5.5.7 Avoidance of Statutory Obligations

Avoidance techniques to get around the Construction Act are now commonplace. Here is an example of a clause aimed at avoiding Section 110(1A).

> The first moiety of retention will become due within twenty eight (28) days of the attendance by the Subcontractor at the **Main Contractor's meeting** prior to Practical Completion (or equivalent) of the Main Contract works. The second moiety of retention will become due within twenty eight (28) days of the attendance by the Subcontractor at the **Main Contractor's meeting** prior to the Certificate of Making Good Defects (or equivalent) under the Main Contract. (emphasis added)

Here the release of retention is made conditional on attendance at the main contractor's meeting (rather than on performance of the main contract obligations or issue of a main contract certificate, both of which would be outlawed by Section 110(1A)).

There seems to be a belief in the industry that one can amend statutory provisions by the simple expedient of inserting a different provision in the contract. Here is an example of one such clause:

> Where the Sub-Contractor intends to exercise its right of suspension for non-payment under Section 112, Housing Grants, Construction and Regeneration Act (as amended) the Sub-Contractor shall issue a notice of suspension at least **42 days** prior to the date on which it intends to suspend. (emphasis added)

The Construction Act only requires that the notice is issued **7 days** prior to suspension.

5.5.8 Not the Whole Story

This is only a small part of the story. The UK construction industry is riven by payment abuse and bullying that, in the worst cases, amount to fraudulent behaviour. Outstanding payments are used to extract or leverage concessions that would not otherwise have been granted. One such example is the case of a specialist contractor that had completed work under a letter of intent. The final payment under the letter of intent would not be made unless the contractor agreed to sign the proffered subcontract. It is also fairly well-established practice to hold back payment due on a final account or not release an outstanding retention unless 10% (or possibly more) is knocked off the final account.

Over the years following the Construction Act reforms and the amendments made to the Act (which were implemented at the end of 2011) both Parliament and governments of various hues have sought to address these issues, particularly where they impacted on public sector construction. For example, in 2002 the Trade and Industry Committee in the House of Commons carried out an inquiry into the practice of retentions. It advised that the practice was 'outdated' and 'unfair' and that they should be phased out in the public sector as soon as possible. This recommendation was echoed in a later report, *Construction Matters*, published by the Commons Business and Enterprise Committee in July 2008. These recommendations were ignored.

5.5.9   Charters and Codes

Many (including governments) have believed that the problem will be solved by voluntary charters and codes. Those that believed this failed to recognise that payment abuse is the result of strongly embedded commercial policies. In 2008 the (then) Government published, to great fanfare, the Prompt Payment Code. Some signatories to this Code have been amongst the worst offenders. One can have a contractual obligation to pay within 150 days and still be within the Code, provided that one did not pay late.

In 2007 the Office of Government Commerce (OGC) – which was then responsible for public sector procurement policy – published its Fair Payment Charter with endorsement from the construction industry. This charter set out eight 'Fair Payment Commitments' including 30 day payment cycles. This model fair payment charter was included in a *Guide to best 'Fair Payment' practices* issued by OGC. Within the list of commitments there was included the following:

- Companies have the right to receive correct full payment as and when due. Deliberate late payment or unjustifiable withholding of payment is ethically not acceptable.
- The correct payment will represent the work properly carried out, or products supplied, in accordance with the contract. Any client arrangements for retention will be replicated on the same contract terms throughout the supply chain....

The Guide went on to recommend the progressive use of PBAs and that evidence of good payment performance should be used by public sector bodies as a key pre-qualifying criterion for the selection of lead contractors. Public sector clients were invited to sign the charter at the outset, while contractors and their suppliers would be expected to sign before appointment.

However, subsequent experience indicated that the charter was either not being signed up to by project participants or, if it was, it was subsequently ignored. The Government Construction Clients Board (GCCB) – which was now in the procurement driving seat – decided to be rather tougher. The Cabinet Office published Procurement Information Note 2/2010, which advised that all payments on government works contracts would be made within 14 days of due payment dates under the main contract and to subcontractors within 19 days of the main contract due dates and to subsubcontractors within 23 days of the main contract due dates.

The Note stated that the GCCB had agreed that central government departments, agencies and non-departmental bodies would move to a position where PBAs were adopted unless there were "compelling reasons not to do so." This still remains the policy of government as a construction client.

Nonetheless it seems that there are some who still never learn from experience. At the beginning of 2013 the Department for Business, Innovation and Skills (as it was then known) was developing a sectoral industrial strategy for construction. This was launched in July 2013 as *Construction 2025*. However, out of nowhere, there appeared in this glossy document a proposal for a Supply Chain Payment Charter. There had been no prior discussion with the industry about this. Again this charter was intended to be voluntary and was officially published in April 2014 by the Construction Leadership Council, a Government appointed body.

The charter had all the usual platitudes including the 'ambition' of getting rid of retentions by 2025. This charter was revised to re-set its targets with the objective that, by the beginning of 2018, everyone would be paid within 30 days. Needless to say this was not achieved. Unfortunately this 'pie in the sky' charter was pedalled by certain industry bodies that either possessed a naivety of child-like proportions or were deliberately seeking to obstruct real improvement.

5.5.10 Supply Chain Finance Initiative

In October 2012 the (then) Prime Minister, David Cameron, announced that his Government would be promoting its Supply Chain Finance Initiative. This was aimed at small enterprises and SMEs. A supplier, on payment of a small fee to the bank nominated by the paying party, would be able to access its cash earlier. However, Cameron didn't anticipate how the construction industry would respond to this. One large company doubled its payment cycles and then invited its suppliers to pay a fee to access their monies within the original payment timescale. Other large companies followed suit and in the process generated large cash bonuses for themselves.

5.5.11 Public Contracts Regulations 2015

In May 2013 Cameron's small business adviser, Lord Young, recommended that payment terms on all public sector contracts and subcontracts should be standardised. The opportunity to implement this came with the introduction of the 2015 Public Contracts Regulations. Regulation 113 obliged public bodies to ensure that 30 day payment clauses were inserted in all contracts, subcontracts and subsubcontracts.

Statutory guidance accompanying Regulation 113 advises that in construction contracts the 30 days should commence from the expiry of the 5 days from the contractual due date for payment. The five days are the statutory period (under the Construction Act) during which either party must issue a payment notice informing the other of the amount of payment considered to be due.

Unfortunately the means of enforcing Regulation 113 is very weak. The absence of a 30 day payment clause in a supply chain contract only gives a contracting party a right to complain to the Mystery Shopper Scheme. This is a complaints process managed by Crown Commercial Services. Unfortunately this scheme lacks enforcement powers. It may be that, in time, such complaints would be dealt with by the recently appointed Small Business Commissioner, who also lacks 'teeth'.

5.5.12 Project Bank Accounts

Reference has already been made to PBAs. In 2016 the Infrastructure and Projects Authority published its *Construction Strategy 2016–2020*. This stated:

> PBAs are recognised as an effective mechanism for facilitating fair payment to the supply chain. Government departments have committed to use PBAs on their projects unless there are compelling reasons not to do so. Improving the financial position of SMEs, reduces the risk of insolvency which can in turn limit the capacity of the market to deliver good value. It can also help to improve the standing of government clients and foster collaborative relationships across the supply chain, leading to increased value for money across programmes of work.

The first PBA in the UK was set up on a Defence Estates (now the Defence Infrastructure Organisation) project at the turn of the century. The project was the Defence Logistics Headquarters at Andover in Hampshire. The Defence Estates required that the contractor, Citex, pay its supply chain as a condition precedent to payment by Defence Estates as the client. Citex – as is the case with many main contractors – did not have the funds to do this. A way around this dilemma was eventually negotiated. Defence Estates would lodge certified payments in a PBA out of which payments would be made directly to Citex's supply chain. During the course of the project Citex went into insolvency. The insolvency practitioner was unsuccessful in obtaining the funds in the PBA for distribution to Citex's creditors.

5.5.13   What is a PBA?

A PBA is a ring-fenced bank account out of which payments are made directly and simultaneously to a lead contractor and members of its supply chain. The PBA must have trust status; monies in the account can only be paid to the beneficiaries – the lead contractor and named supply chain members. The bank account should be held in the names of the trustees, who are likely to be the client and lead contractor (but members of the supply chain could also be trustees).

The advantage of trust status is that an insolvency practitioner (administrator or liquidator) appointed to the lead contractor's company cannot have access to the monies in the account. This is because the monies are 'owned' by the beneficiaries of the trust, that is, both the lead contractor and members of the supply chain. The motivation for some clients using PBAs is that they minimise disruption when a lead contractor goes into insolvency.

A PBA may be suitable for a £1 million project or even smaller projects. At the beginning of 2013 the Northern Ireland government announced that PBAs would be used on public sector projects over £1 m although this was subsequently raised to £2 million. For projects of short duration, such as a few weeks, PBAs may not be necessary.

The benefits of having a PBA in place outweigh most other options for achieving fair payment (apart from direct payments from a client to the supply chain):

- Once the monies are deposited in the PBA the supply chain's cash is safe since the PBA is ring-fenced.
- Payments out of the PBA are made simultaneously to everybody; payments become more efficient and cash flow management more transparent.
- Payment times are the same for the main contractor and firms in the supply chain.
- Clients could also make savings since financing charges across the supply chain are reduced.
- There is less risk to clients of disruption and delay as a result of the supply chain suspending work for non-payment.
- PBAs increase collaboration since there is more trust; the delivery team can concentrate on working together without the distraction of payment delays and abuse.
  Since the lead contractor doesn't have access to its supply chain's cash the temptation to use it isn't there. This also applies downstream to tier-two and tier-three contractors. Thirty day payments will often end up being ninety day payments and longer by the time they get to the end of the supply chain. Table 5.1 illustrates this.

Table 5.1

Traditional payment arrangements		Project bank account	
Days		Days	
Client to contractor	30	Client to PBA	30
Contractor to subcontractor (usually much more)	30	Contractor and subcontractors	5[a]
Subcontractor to subsubcontractor or supplier (usually much more)	30		35
	90		

a) This is the maximum length of time for the money to be in the PBA.

From the above it can be seen that a PBA removes a long-winded, potentially abusive and hierarchical payment structure.

PBAs do not cut across the contractual provisions governing entitlement to and discharge of payments. They are simply a safe receptacle for the cash. Once the lead contractor has agreed its application with the client the agreed amount is paid into the PBA. The contractor then issues the authority to the bank listing the names of the recipients and the individual amounts to be paid. The client will also sign off the authority. All payments in and out of the account are made electronically, which means that monies should not be in the account for more than five days. If, for whatever reason, there is a shortfall in the account, the lead contractor remains bound to pay what is due to its supply chain members. **This is not a pay when paid mechanism.**

> *Without doubt this has to be the way forward for our industry if the current payment culture is to be eradicated. PBAs help eliminate the excuses for late or reduced payment, the burdens on overhead costs and the programme delays as a result of disputes and resultant insolvencies, which often result in SMEs being hardest hit.*
>
> Director of an SME mechanical and electrical firm

To date no alternative to PBAs has been offered as providing a more effective solution to the persistent and endemic problem of late and non-payment in construction. It does behove those who are critical of the concept to develop solutions that will have at least the same impact as PBAs.

5.5.14 Retentions

The practice of retentions has been in existence for almost 200 years. It was originally introduced to provide a measure of security in the event that a contractor went into insolvency. Today it is claimed that retentions are necessary to ensure that a contractor returns to remedy non-compliant work within a so-called Defects Liability Period, although industry standard contracts do not define with any precision the reason for which retentions are withheld.

Research carried out for the UK government in 2017 indicated that the prime motive for withholding retentions is to boost the cash flow of the party deducting them. This even applies to many public bodies that require retentions to finance other works or activities.

The abuse associated with the practice is well known. Retentions are rarely released on time and in many cases are never released. Many small firms give up chasing the monies. This was what Her Majesty's Revenue and Customs had to say:

> In recent years, construction industry customers have become increasingly reluctant to pay retention monies, irrespective of whether there are defects to be made good. It is now common for such monies never to get paid.

Cash retentions also amount to a subsidy to financially strapped businesses in the industry. They subsidise incompetently run businesses. This was acknowledged by the New Zealand Government when recently passing legislation that required that retentions were to be placed in trust.

> *The use of retentions as working capital is the key concern for the Government; funding working capital from retentions can mask and reward poor performance and poor financial management practices. For example, undercapitalisation and low-price tendering are long standing features of the Construction market that contribute to its low productivity and innovation. The use of retentions as working capital enables those features to remain with no incentive to change and no incentive for clients or lead contractors to properly manage project risks.*
>
> New Zealand Government Ministry of Business, Innovation and Employment.

Retentions are also the antithesis of trust. It is a strange way to kick off a commercial relationship by suggesting to a contractor that you don't trust it to properly complete the job and therefore you are holding back 5% from its due payments. There is so much pre-qualifying, accrediting and auditing of firms today that one would think that this would provide a measure of comfort when seeking a competent firm.

On the other hand there are no barriers to entry to the industry. The ability to offer the lowest price is often the dominant criteria in selection. No doubt some would view retentions as a small cushion should the firm that has offered the lowest price goes into insolvency or simply disappears.

5.5.15   The Cost of the Retention System

Research commissioned by the Business Department in 2015 and published two years later in October 2017 revealed that £700 m of cash retentions (at 2016 prices) had been lost as a result of upstream insolvencies. This represents almost £900 000 lost each working day by the industry. Retention monies are legally owned by the party from whom they have been withheld. Consent to their withholding doesn't extend to their being made available to satisfy the creditors of the other party in the event of that party's insolvency.

This lack of security for retention monies also means that firms cannot, in turn, use their retention ledgers as security to increase their borrowing facilities. However, there are other costs. The bulk of these are expended by firms in chasing outstanding retentions; often they are not released until some years after handover. Some tier-one contractors are known to release only a small percentage of outstanding retentions at handover (less than the standard 50%) and release the balance some two or three years after handover.

The late release of retentions also deprives firms of millions of pounds in interest. Whilst there exists a statutory right to interest for late payment, firms in the industry are often too commercially afraid to use it.

So what's to be done?

In *Constructing the Team* Latham recommended that cash retentions must be placed in trust. This never happened.

Whilst some in the industry will call for a statutory ban on withholding retentions there is likely to be considerable opposition to this. However, one state in the United States, New Mexico, has already outlawed retentions. Article 28 of the 2013 New Mexico Statutes states (Section 57-28-5):

> When making payments an owner, contractor or subcontractor shall not retain, withhold, hold back or in any other manner not pay amounts owed for work performed.

This provision originally came into force in 2007 and was consolidated into the 2013 New Mexico statutes.

The easier way forward is to ring-fence cash retentions in the way advocated by Latham. Using an existing model is always the better option. One such model is the protection offered to shorthold tenants under Section 215 of the Housing Act 2014. This requires that a landlord requiring a cash deposit from a tenant as security for the performance of the latter's obligations under the tenancy agreement must deposit it in a government approved scheme. To date there are three such schemes.

Equally legislation could provide that cash retentions are deposited in a retention deposit scheme. This would ensure that the monies are safe from upstream insolvencies and that they would be released on time.

5.6   The Future

UK construction is at a major crossroads. The UK's exit from the European Union is likely to have a significant impact on its cost base. Both public and private sector clients are increasing pressure on the industry to innovate, primarily through investment in digital and manufacturing technologies. The industry is being driven to make greater use of robotics, augmented reality tools and 3D printing.

These demands and pressures are unlikely to be fully addressed unless the industry's inefficient payment practices are dragged out of the nineteenth century into the twenty-first century. This will mean a radical change in the business models of many companies, particularly the largest contractors. The use of digital tools such as blockchain technologies will need to come to the fore to create greater transparency in project payment systems, thus ensuring that cash flow is being constantly maintained to all participants in the delivery process.

In December 1993 Latham published an interim report titled *Trust and Money*. In that report he observed that instead of oil (money) the construction engine had grit in it. The engine still has grit and, whilst it remains, the consequent lack of trust in the industry will continue to obstruct the collaborative effort that is desperately needed to bring the UK construction delivery processes into the modern era.

6

A Discussion on Preventing Corporate Failure: Learning from the UK Construction Crisis

Stephen Woodward[1] and Nigel Brindley[2]

[1]*Corporate Risk Management Consultant*
[2]*Director Infraquest Limited*

6.1   A Call to Action'

[1]The global [2]construction industry can learn lessons from UK corporate failures. In doing so the risk profile of infrastructure should improve as an asset class, making it less risky for investors.

6.1.1   Corporate Risk Manager

Risk profile: the demise of the UK's second largest contractor[3] changed the horizon. What was happening before resembled arranging deck chairs on the Titanic. The industry rightly acclaimed belated progress in collaboration, building information modelling (BIM), and improved forms of contract. However the iceberg looms in the form of:

- Tier-one contractor insolvency
- Supply chain company destruction
- Risk pass-down to those least able to bear it
- Unwavering focus on price rather than qualitative value
- Large revenues unbacked by corresponding balance sheet strength.

  Brokenness of the model: what is happening in the UK could happen in other countries; however, there are essential differences between the UK and other markets. Why

1 Stephen Woodward, for two decades a construction risk advisor to funders and investors (including UK government office) and Nigel Brindley, infrastructure investment banker, fund manager, and non-executive director, have both been directors at international contracting organisations during their respective careers.
2 Stephen Woodward, corporate risk manager, and Nigel Brindley investment banker, discuss the viability of tier-one contractors and supply chains in respect of the UK construction crisis and the lessons learned for the global construction industry in improving the risk profile of infrastructure as an asset class by making construction less risky for investors.
3 *Carillion plc was a multinational facilities management and construction services company listed on the London Stock Exchange. The company experienced financial difficulties in 2017 and went into compulsory liquidation on 15 January 2018. In April 2018 the Official Receiver estimated its liabilities at £6.9 billion, a figure over three times higher than given in the accounts at the end of 2016. Source: Daily Telegraph 14th April 2018.*

is it that the major contractors of France and Spain are larger and stronger than those of the UK? Their approach to public sector procurement, government patronage, quality of executive management and cut-throat competition within a fragmented market are all distinctive differences. The increasingly asked questions relate to who is next and just how safe is any contractor? Can the construction industry survive in its present form or are radical and urgent changes needed to improve management accountability, risk management, cash flow, and profitability?

6.1.2 Investment Banker

Financial control and resourcing: contractors clearly show an inability to ride the economic cycle. The global industry norm is project gestation periods of two–three years with pricing at one point and construction at another point in the cycle. Without a contractual rise and fall mechanism, pricing in one cycle and delivering in another has dramatic effects in construction the world over. Then there are the ever-present knock-on effects of 'problem' projects. The dominant culture is low bidding and low margins with 'clashes' arising from differing international and corporate cultures, bureaucracies, and administrations. Problem projects quickly erode corporate margins as resources become stretched and time extended. Resources are then unavailable for follow-on projects. This creates a vicious circle. Risk pricing is notoriously difficult and construction risk seldom correlates with general inflation. Widely differing regional economic variations, swings in global commodity pricing, and skills shortages enhance risk.

We believe that there are essential differences in margins, contractual risk apportionments, and volatility and visibility of infrastructure and public sector construction programmes between national markets.

Globalisation and digitalisation: there is now something more serious at play over and above the normal 'boom and bust' cycle. This is the rapidly changing global construction business environment.

- Several tier-one contractors are in serious financial trouble globally, but are continuing to operate. This is against a background of takeovers to save companies, corporate disposals by investors trying to recover stock market losses and **overtrading** caused by acquisition financing.
- Within an environment of intense competition untenably low margins using **zero margin bidding** are all too common. This causes structural corporate trading difficulties stemming from lack of realistic contingency in pricing contracts. Endemic low margins constrain overhead spend. Any increase in overheads adversely affects critical investment shortfall in systems, skills, and emerging technology.
- There is too little importance ascribed to cash flow. **Revenue is the key driver**. Profit declaration takes place before cash generation or work is sufficiently progressed with overstatement of project profits and problems rolled forward to support future revenue through deal flow. This allows offsetting of current losses against prematurely declared future profits with corporate focus on revenue rather than cash flow and profit.
- To fight cash flow problems tier-one contractors mitigate **liquidity risk** to solve cash flow problems by exploiting supply chain contractors (stretching supplier and subcontractor payment terms) and using them as a bank. This leaves the supply chain

underfunded through extended payment terms, undervaluing completed work and violating retention accountability. Payments taking up to 120 days, without taking into consideration payment disputes and withholding of retention monies, are clearly unsustainable. If a supply chain contractor looks in trouble, instead of helping by prompt payment, delays take place thus worsening cash flow by withholding even more monies.

- Funds from acquisitions are **concealing financial reality** with acquisitions burying existing risks. Acquisition financing helps disguise deep-routed financial management problems. Larger than necessary reserves can be created against a 'bought' balance sheet and are offset against over-generous goodwill allowances. A cycle of repeated acquisitions enables serial application of the practice…until the cycle stops!

- The prevailing **macho culture** suppresses risk recognition and risk allocation. The payment system is tantamount to corporate bullying, pushing risks disproportionality down the supply chain to those who can least afford to bear them. This is artificial risk apportionment as most of the liability associated with these risks reverts back up the supply chain in the event of supplier insolvency. There is too much buying power concentrated in too few large tier-one contractors. This creates disproportionate commercial power. In the hierarchical culture, supply chain contractors tend to accept too much risk in fear of losing to bidders who give unqualified Bids

- **Internal management** often delay or under-estimate profit exposure from project losses. There is a culture of 'deliver the result and no questions will be asked'. A business manager is more inclined to disguise profit challenges in the hope of trading through them than face career limiting early disclosures. Commonly, this culture pervades throughout the management structure.

- There is often **a disconnect between finance and operational functions**. Consequently, there can be too little communication between them and too little appreciation of what the other does. Often, project management and finance systems are not integrated thereby giving different perceptions of operational profitability. An example is the lack of insight of the finance function into the operational problems.

- On projects needing contractor investment, the practice of earning success fees on financial close of multi-million pound schemes distorts normal commercial checks and balances. Closing a Bid secures cash now with instant P&L contribution, while not doing so crystallises Bid costs with consequential negative profit and loss impact. Commercial, technical, financial, and legal advisers are all motivated to support a Bid as significant proportions of their fees can be success related.

6.2 Lifting the General Level of Corporate Management

Corporate Risk Manager

Will construction contractors continue to hit the iceberg? This depends on the willingness and speed of response by the industry to follow a new course and changing its approach to corporate management, including much more rigorous risk management. It is abundantly obvious from the performance of contractors that something is fundamentally wrong with UK construction and there is an increasingly held view that the contracting model no longer works in its traditional form. If top contractors diligently used more effective systemic risk management they would become even more attractive to investors, be it financial, or those who invest themselves through employment, as direct suppliers, or subcontractors, or through the supply chain.

The Farmer Review has an apt subtitle: *Modernise or Die – Time to decide the industry's future*. The current crisis follows two earlier government reports into industry practice: *Constructing the Team* (Sir Michael Latham, 1994) and *Rethinking Construction* (Sir John Egan, 1998). The adversarial culture within the construction industry means that identification, sharing, allocation, and management of risk is extremely poor, and risk management as a systemic discipline is not prioritised in project and corporate planning and delivery. This starts with clients (including the public sector), who are too willing to delegate risk into the supply chain inappropriately and prioritise lowest price over qualitative assessment and value for money. The tier-one contractors recognise this trait, which encourages cost cutting and risk taking as the only sustainable strategy to contract wins.

The cyclical corporate failure risk is good client value (including for the public sector); for the rest of the time the supply chain absorbs the risk and the cost of contractor failures tends not to revert to client accounts. The prevailing culture results in risk passing down the supply chain and coming back up as conflict.

Confidential interviews with industry leaders show chronic failure in the following key areas:

- *Return on investment*. To compensate for 'on-project' margin erosion the motivation is for contractors to 'save-up' potential claims for settlement at the end of the project – *'claims culture'*.
- *'Balance Sheet Light'*. Construction groups have very light balance sheets compared to similar turnover businesses in other sectors. Plant and equipment tends to be hired against specific contracts or purchased on finance leases. This flatters a construction company's return on investment and return on capital employed ratios and accentuates their financial volatility.
- *Risk pass down*. Contractors forced to accept risks disproportionate to their competences and margins naturally focus on corporate protection rather than collaboration – *'silently pass the risk on'*.
- *Work-winning momentum*. Overriding risk assessment to chase turnover – *'we must win this job!'*
- *Culture*. Collaboration is an anathema to an industry built on an adversarial culture – *'why collaborate when you are not getting paid?'* The only route to profit is to aggressively pursue claims at the end of a contract, at which stage the contractor knows the claim value needed to recover on-project losses.
- *Reporting*. Reports saying what the recipient wants to hear, not – *'don't' do the deal!'*
- *Optimism bias*. An untested personal belief that events will come right – *'the next project will turn out better than the last'*.
- *'Trading the future'*. Settle the claims on this contract under value in return for the opportunity to Bid on the next contract.
- *Work in progress*. Adjustments to give better than true figures – *'it really is better than these figures show'*. Unrealistic over-valuation of variations, and delay and disruption claims to meet profit targets.

From interviews the most commonly identified management failures are:

6.2.1 Operational

- Inadequate management experience
- Lack of technical bidding capability
- Lack of essential project delivery ability
- Management over-stretch (trying to do too much with too little resource)
- Lack of investment in systems and skills
- Specification scope creep
- Undeliverable output specifications
- Unfamiliar operational issues
- **Human conflict arising from poor behavioural skills and differing working cultures**

6.2.2 Commercial

- Chasing corporate revenue by bidding (near) zero-margin and hoping to obtain margin during project delivery
- Under-pricing
- Budget scope creep
- Lack of commercial project management skills
- Inability to track contract changes, delays, and disruption in real time
- Using costs-to-date rather than costs-to-complete with budget comparison
- Over-optimism of contract outcomes
- Unrealistic valuing of un-submitted claims
- Taking value of not yet finally agreed claims
- **Lack of transparency, openness, and honesty**

6.2.3 Financial

- Insufficient working capital
- Failure to accurately value long-term work in progress
- A culture of financial engineering to show a better than actual picture
- Lack of robustness and accuracy of auditors reporting
- Unfit for purpose accounting and reporting processes and standards

Investment Banker

- It is a never-ending cycle. A corporate team is the same as a project team; they just scale up the problem and filter the messages as they see fit. 'Soft' issues are the real problem.
- Against the background of poor corporate governance there is very little cross fertilisation of staff and skills in construction as in other industries. Because of its projects culture, staff promotions often go beyond levels of trained competence and ability. This gives rise to the further risk of replication of problems from a lack of knowledge as to what management processes work and how to avoid pitfalls. With a lack of continuous improvement the same mistakes are repeated.

- The low entry barriers ensure a plentiful supply of low-cost competition. This maintains a constant downward pressure on contract pricing. At the same time, as contract liabilities tend to lag revenues (generally, contractors pay suppliers after receiving payment for the corresponding work themselves), tier-two and tier-three contractors are constantly hungry for cash flow. Consequently, many smaller subcontractors would rather take any profit today at the risk of financial distress tomorrow in order to stave off the risk of financial distress today.

  The issue here is that a small subcontractor is reliant on the tier-one contractor(s) with which it has a relationship for its business. It will accept whatever margin it can secure because the lost business cannot be quickly replaced, and without it many will face short-term insolvency. Tier-one contractors feed on this; low entry barriers in general contracting mean that there are plenty of alternatives ready to step into 'dead men's shoes'. Small and medium sized companies have a problem because they suffer as much, or often more, than large companies, do not get the publicity, and more importantly do not have the same ability as larger companies to raise capital to dig their way out of trouble.

- The value of any company is the value of its future cash flows, with contractors motivated to prove their strength through their order book and associated projected cash generation at (seldom achievable) Bid margins. Current losses are a charge against over-hyped future revenues (that cannot be disproved) and the cycle repeats.

- Part of the problem is that entry barriers at trade level are low. Consequently, the industry endures a high level of insolvency without undue compromise to work flow. The larger contractors exploit this tendency and withhold supply chain cash under the premise of a dispute and then benefit from a supply chain insolvency. Smaller subcontractors and suppliers also exploit the trend by accepting insolvency today and re-starting a new company tomorrow (there is little or no IP or asset requirement to hinder this – 'man in a van syndrome'). If there is a pending insolvency suspected, many tier-one contractors hold on to cash under the premise of incurring added costs in sourcing an alternative company and recognition of liabilities that can no longer be passed down).

6.3   Improving Risk Management

Corporate Risk Manager

There is failure in effective 'fit-for-purpose' risk management at both the corporate and project levels.

At corporate level there is compliance failure with the requirements of the Financial Reporting Council (FRC) in setting high quality corporate governance and stewardship codes for financial and investment reporting.

The FRC publishes *Guidance on Risk Management, Internal Control and Related Financial and Business Reporting*. The guidance recommends that corporate governance defines the process adopted for ongoing risk monitoring and review, including risk reporting and the likelihood, impact, and ability to respond to change.

FRC Guidance (in Summary)
- Ongoing monitoring and review of risk management and internal control systems with verifiable proof that systems align with strategic aims and deliver effective internal control of risks

Asks Direct Questions
- *What are your processes to check the effective application of systems for risk management and internal control? How do monitoring and review processes effectively re-evaluate risks and adjust controls in response to changing goals, business, and external environment? How are they adjusted to reflect new and changing risks, operational deficiencies, and emerging risks?*
- *What are your channels of communication that enable individuals, including third parties, to report concerns?*
- *How is inappropriate behaviour dealt with? Does this present consequential risk?*

Investment Banker

- Programmes and projects are the output of contracting.
- It is vital for investor confidence that the process of construction procurement has truly effective risk management. The problem is that there is little appreciation among financial investors of the nature of construction risk and consequentially they poorly manage investment risk exposure.
- Lessons learned from construction output must inform corporate risk management so that exemplar corporate risk management is the top of the pyramid. Risk management then cascades from corporate to programmes and projects, and is constantly evolving, renewing, and innovating to meet the demands of a fast-changing world.

Corporate Risk Manager

The four components needed to achieve better results from the risk register and use of systemic risk management are:
- Early risk identification
- Properly considered risk sharing and ownership
- The direct linking of risk ownership and reward
- Continuous risk feedback and evaluation through effective monitoring

6.4 Joint Recommendations by the Corporate Risk Manager and the Investment Banker

This translates into using systemic risk management and "mining" the best information available to support systemic risk management.

6.4.1 Programme and/or Project Risk Register Actions

- Establish **risk management primacy** to give the best opportunity for cost and time certainty.
- Use risk management to create the greatest potential for **collaborative working** by aligning goals and certainty of outcomes.
- Embed the risk register within the programme and/or project **business case** as the level-one control document. The positive effects will then flow through the programme and/or project, including the supply chain.
- Actively **engage stakeholders**, collect the best information available and give feedback to make the risk register an integrated, effective, and engaging risk management tool. A passive risk register doesn't work! Remember the adage of 'rubbish in, rubbish out'.
- Use **proprietary early warning systems** for gathering risk information to inform the risk register by engaging with all stakeholders, mining unbiased information, and generating action focused feedback reports to form realistic risk mitigation plans.

6.4.2 Gateway Reviews (GRs)

- A best practice example of **systemic risk management** for a project or programme is a gateway process.
- One step at a time! The simple rule is do not go from one stage to the next until there is robust testing for completeness, comprehensive auditing, and certainty that all risks have been identified before signing off on the stage.
  The Office of Government Commerce (OGC) Gateway ™ is an exemplar GR process:
- Part of the Infrastructure and Projects Authority (IPA) toolkit
- One of the tools for public sector procurement assurance reviews
- An independent process covering the complete project cycle
- Confidential reviews at the critical stages of a programme and/or project:
  - Strategic assessment
  - Business justification
  - Delivery strategy
  - Investment decision
  - Readiness for service
  - Operations review.

In addition to a peer review from one stage to the next, a gateway process ensures the embedding of risk management within the business case, supports preparation of a risk register, and the creation of a risk profile on day one of the project. It also ensures monitoring variance to the set risk profile forces:

- Transparency in risk identification to achieve fair risk sharing and in particular avoiding unfairly loading risk down the supply chain during construction tendering.
- The party best able to manage the risk holds the risk.

As with the preparation of the risk register for the principle programme and/ or project control, any gateway process is only as good as the integrity of the confidential reviews and the monitoring of the set risk profile. The key to success is information gathering and feedback to act as evidence of progress and early warning of a divergence from plan.

6.4.3   Early Warning Systems (EWS)

As a tool of best practice risk management EWSs are necessary for:
- Best quality information gathering and feedback for the risk register and systemic risk management use.
- High-level reporting to stakeholders, including principals, decision-makers, and investors.
- Making stakeholder engagement the most effective means of top-down change management.
- Culture change – in a hierarchy EWSs support corporate risk management and programme/project risk management and are supported themselves by (i) unbiased information gathering and reporting and (ii) identification, mapping and engagement of stakeholders.[4]
- Supporting procurement. The New Engineering Contract pioneered the concept of EWSs to improve communication and reduce conflict. Systems suppliers now need to do more in exploiting the commercial opportunity for proprietary systems to support and move the EWS concept forward.

Developing innovative risk tools – A typical first mover proprietary EWS is the RADAR system from ResoLex UK featured in *Project*, the journal of the Association for Project Management Spring Issue 2018. ResoLex subsequently collaborated with the project manager 3PM and Professor John French at the University of East Anglia to produce a research paper, which is hosted at www http://www.resolex.com/casestudy-UEA-TEC analysing this system's use as an innovative risk management tool.

6.5   Conclusions

- Industry must tackle head-on the risks and opportunities of globalisation and digitalisation.
- Only efficient and well managed contractors are profitable.
- Contracting models urgently need restructuring.
- Currently there is little left for reinvestment in training and technology.
- Construction needs de-risking in order to make it a better infrastructure asset class. The problem here is that investors in projects benefit from a tier-one contractor wrap that only proves worthless once in a generation (Carillion this decade, Jarvis last decade, etc.) – this rate of failure justifies investors being content with the current system for gains todays versus once-in-a-generation failure, which they will argue is covered by portfolio risk.
- There should be tender compliance whereby bidders must demonstrate that they have used robust corporate and project risk management processes in their Bid preparation.
- The public sector must lead by example in openly expressing upfront Bid evaluation to show criteria and weightings for evaluating tenders in addition to cost rather than pursuing 'cost-based' evaluation of tenders.

---

4 Refer to the concepts of 'the logical levels of organization in systems' developed by Robert Dilts in his book *Changing Belief Systems with NLP*.

- Set up prequalification compliance requirements compelling bidders for public projects to declare construction risks of work in progress in terms of project and contract value by sector and region in relation to the size of the overall business.
- The public sector should mandate that all contract enquiries include a risk apportionment matrix that realistically apportions risk to the party best placed to manage it.
- Financial regulations need changing to restrict the abuse of financial engineering to show a rosier picture.
- Better information for investors needs to be provided, and they need to be given the opportunity to influence corporate outcomes through more effective open and transparent reporting of risk and reward.

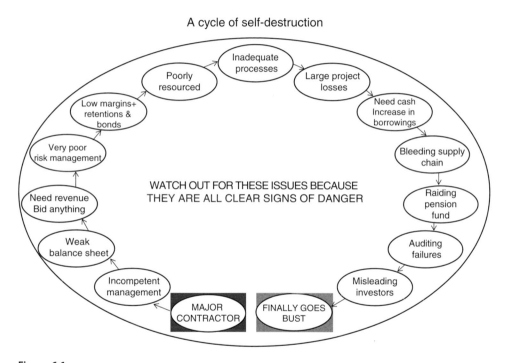

Figure 6.1

Section B – People and Teamwork

7

Obstacles to Senior Management and Board Success

7.1 Introduction

This chapter is the outcome of conversations with current and former senior personnel within the industry, and psychologists specialising in *'team dynamics'* and behavioural observation.

In discussions about team dynamics psychologists talk about the need to find a balance of individuals with different colour personalities – *'red'* being assertive, even dominating; 'blue' the detailed perfectionist; 'green' the calming mediator; and 'yellow' more of a laid-back enthusiast. Effective tcams will normally comprise an appropriate balance of personality colours – except, as one psychologist remarked very pointedly, *'in your industry (construction), where such a balance is not apparent, and nearly everyone in senior management is most definitely "red"'!*

This observation is one that has been commented upon many times over the years by all sorts of people connected with the industry. Certainly, the broad construction industry management style has clearly been characterised as *'be "red" or dead'*. Nevertheless, some progress has been made, particularly with the development and successful implementation of more collaborative forms of contract and working agreements. However, when you consider dispassionately the way in which the industry still creates and manages its senior leadership teams, it remains apparent that this process still often occurs 'by default', lacks proper objectivity, and too often fails to focus on developing a management structure that can optimise team and leadership performance.

Surely, for shareholders and for the business, such optimisation must be a key objective? In a sector that is all about managing high risks (for pretty minimal reward in comparison with many other industries), the quality and effectiveness of the senior management team are key in how a business manages and protects itself from inherent hazards. By 'management', we refer not only to the executive management, responsible for the day-to-day running of the business, but also to the formal Board of the company (including non-executives) and its role in providing independent oversight and governance.

So what appear to be the most common areas of senior management failure in construction sector businesses, and why? These are discussed below, in no particular order. Indeed, all of the factors described are to some extent both interrelated and interdependent.

Global Construction Success, First Edition. Charles O'Neil.
© 2019 John Wiley & Sons Ltd. Published 2019 by John Wiley & Sons Ltd.

7.2 Groupthink and Team Selection

How often have we all witnessed a senior manager struggling to cope with the demands of their promoted role? Construction seems to have a fatalistic assumption that good project engineers will automatically make good project managers, who then make good business unit managers, who will then easily convert to make good senior managers or Board-level directors? Such thinking has three direct and unfortunate consequences:

- The company will lose good people. The over-promoted senior project manager will leave because they fail to transition into being a competent and/or content business unit leader. The company has then lost itself a good senior project manager, primarily because senior management fails to recognise both the limitations and the real skills of the individual. Conversely, the individual with the diversity and outlook to become a strong business leader may leave because they are not a particularly strong project manager, and may feel that their chances of promotion to senior management level within the company are limited because of this.
- Senior management will tend to fill its ranks with individuals of similar characteristics. Promotion becomes a matter of choosing individuals with common traits, outlook and experiences as the existing management. This is one primary reason why senior management in construction has too many similar 'red' personalities. We have probably all been guilty of this in our own careers and there are some lessons to be learnt from listening to the thoughts of one of our friendly psychologists. Sitting with an interview panel to select a new business leader, the psychologist was asked to independently consider both the composition of the panel and the candidates. Even before the interviews the psychologist correctly ranked the order in which the panel would subsequently mark the candidates (and why), taking into account how they would rank candidates with regard to the needs for the management composition and leadership balance for the business unit; then saying *'You have too much "red" assertiveness and need some "green" mediation skills to counteract this'* and recommended a particular candidate. The psychologist was so persuasive that the panel eventually ran with the recommended 'green' candidate (rather than one of the two 'red' candidates the panel had focused on); this turned out to be an extremely good management appointment, giving much better balance and long-term effectiveness to that business.
- Having management (or a Board) with similar personality characteristics and/or skills promotes groupthink and provides little diversity (or incentive) to ask different or difficult questions, to propose alternative strategies and actions, or to challenge the status quo. As well as being an opportunity lost for diversification and development of the business, groupthink can also be very dangerous – unified thinking leading the company into poor decision-making through tunnel vision in its processes, prejudices and outlook. At Board level this is also a significant issue – any Board dominated by engineers (or accountants, or any other specialist field for that matter) is much less effective in its decision-making outcomes than the Board that encompasses a broad range of relevant skills, experiences and personalities. It was Mark Twain who once observed that, *'when you find yourself on the side of the majority, then you should always pause and reflect'*.

7.3 Training

There is nothing wrong with specialism. However, as an individual's career progresses to senior management and Board-level situations, then the contribution and effectiveness of that person can be greatly enhanced by the personal diversity of experience and skills that they can bring. Although the situation has improved in the past 20 years, we still observe that the training of our construction professionals is largely within quite narrowly defined pathways (and with quite a lot of 'groupthink' in the training methodology). There is also a tendency to *'box'* people in the skills that they have (successfully) demonstrated in their work to date. So, employee A becomes the person to always lead the crisis management team for problem type X, and employee B will find themselves pigeon-holed as the key person in seeking commercial resolution of claim type Y, etc. Whilst it is a natural response to focus individuals towards continuing to 'do what they are good at', this prioritises short term business demands over the need to develop individuals with skills diversity that would enhance their value in future senior leadership roles.

It is also often overlooked by senior management and Boards that they have direct accountability to develop future management and leadership teams as a key part of their ongoing responsibilities. This needs to be about more than sending high-potential staff on the occasional (or even regular) construction industry training course(s) – it is about fostering and encouraging wider personal experiences that can provide future leadership with the skills to deliver innovative and competitive advantage. It is also about recognising the differences in style, process and oversight that are required by individuals as their seniority increases within the organisation.

So, what should we be seeking to develop towards, and what should we realistically expect of our key staff in senior management positions? Perhaps:

- That they have demonstrated at least one core competence in their career (be it engineering, financial, commercial, or whatever).
- That they have then proactively expanded their thinking and knowledge throughout their career path to observe and absorb other competencies (as an example, perhaps every senior manager at executive or Board level in construction should have a basic and appropriate understanding of some contract law, of accounting principles, of communication skills, or perhaps even psychology, etc.?).
- Over time, and through a combination of direct experience and appropriate training (see below), to have established a basic competence in information gathering, risk analysis and strategic decision making. These are concepts that can only be taught to a certain level – at some point there becomes no substitute for direct experience.
- They should broaden their minds and properly listen to others (a very underrated and underused skill, in most industries), being able to rationally articulate and convince others, and to understand both their own personal limitations and also the value contributions that they can realistically make as part of a senior management team. This includes a mature understanding of team dynamics and how to use this effectively, as well as being sufficiently broad minded to see 'the bigger picture'.
- Of course, not everyone can know every single thing, and knowledge acquisition is a continual process. The most effective team (or business) leaders over the long term

are generally proven to be those who can recognise and react to this reality – they accept their own limitations, and are not afraid to surround themselves with skilled people, and also to accept challenge from them. Training for strong management and leadership is about so much more than merely counting up the number of formal training courses attended. As Albert Einstein once remarked, '*Learning is experience. Everything else is just information*'!

7.4 Choosing the Wrong Strategy and/or Projects

Reflecting back on their own experiences, some colleagues have considered their highest stress levels to have been encountered when working on projects in their 20s and early 30s, particularly on smaller projects. There were just so many day-to-day decisions to be made, and they survived very much on their own instincts and in building a strong core team around themselves. Many of these decisions had to be made quickly, and often alone. The consequences of making the wrong decision (which inevitably did happen now and again) may occasionally have been major for the immediate project, but at that level they rarely threatened catastrophe for the overall business. As careers develop there will inevitably be an increasingly wide support network, and much greater advisory or analysis resources will become accessible to help decision making. More and more time should also become available to consider, analyse, and make informed decisions. Generally this is a good thing, as the more senior you become in the decision-making process, then the greater is the overall impact and materiality of the decisions being made (either individually, or increasingly as part of a senior management team or Board). In all parts of the construction industry, the consequences of adopting the wrong strategy or in pursuing 'inappropriate' projects can very rapidly become fatal to that company. So many 'catastrophe' strategies or projects can be traced directly back to poor decision making at the initiation or decision-making stages. Not only can these examples (many of which are well described elsewhere in this book) be financially ruinous, but the amount of senior management time then required over an extended period in order to address the consequences will always distract leadership from its other key responsibilities and duties, including neglecting potential business opportunities through lack of available time and resource to pursue them.

Accordingly, one of the key personal transitions for anyone evolving into a senior management role is to develop refined and changed skills in their decision-making techniques. There is no longer the need for frequent, intense and time-pressured direction, and learning to listen, to question and to utilise the wider availability of resources will permit full and informed consideration of all knowledge to then provide timely direction.

Also important is the increasing need to consider a wider corporate picture, encompassing consequence, accountability and governance, which all form part of the collective responsibility discharged by a senior management group or a Board.

It was Professor Michael Porter who once declared that, '*The essence of strategy is choosing what not to do*'. Undoubtedly, the toughest business decision at senior level is always the one that requires you to say '*no*' – to clients, to colleagues, to partners, or whomever. Because the consequences and outcomes of key senior management and Board decisions can be so critical, then it is necessary for the senior individuals within these bodies to develop an appropriate level of objectivity and dispassion in their deliberation. It is not always possible (or desirable) to fully explain to outsiders the basis behind a particular collective decision by a management team or Board. As a leader your emotional commitment to the business (including towards your staff, clients, etc.) should continue, but also needs to change in its nature, introducing a certain level of dispassion. The decisions made are increasingly for the wider business interest, and less to satisfy individual stakeholders within it. Many managers struggle to accept and implement this fact as they progress upwards in their careers. As General Colin Powell once put it, *'Being responsible sometimes means pissing people off'*.

7.5 Need for 'Macro-Level' Focus, with Effective Corporate Oversight ('the Wider Picture')

In a senior management or Board role, *'knowing all of the details'* is considerably less important than *'knowing all of the important details'*. Accordingly, as careers develop, a key management skill is characterised by increasing movement from the strong micro-management control of successful project and business unit management, towards the ability to objectively (and often swiftly) assess a wide range of information, and to then efficiently determine which are the *'important'* parts? Determining correctly *'what exactly is it that I (or we) have to decide?'* is often no less important than making the actual decision. There is no real value in having a clear and precise answer to an imperfect question! Successful senior management and Boards focus on analysing, identifying, and concentrating primarily on those risks and/or decisions that have the greatest impact at a corporate or strategic level. It is an inability to transition from micro-control into this different level and style of macro-management that is very often a key failure factor where good mid-level managers prove themselves unable to make the grade at senior corporate level.

A senior management or Board that fails to implement effective corporate oversight will fail in its responsibilities. That oversight has such a bad reputation is probably unsurprising. We can all recall experiences of working in middle or lower management and cursing the 'interference' of head office. *'They don't understand the way the business works on the ground'* or *'they're all a bunch of jumped-up administrators'* are the cries too often heard around the time of a departmental visit from a centralised team. However, it need not and should not be so. Oversight and corporate governance are both necessary and, when appropriately implemented, are valuable components of effective management at senior level. They are a further area in which the individual manager needs to educate themselves as they rise through the company structure. This forms another important component of being able to develop a wider corporate perspective for senior level decision making. Key success factors in implementing strong corporate oversight include having senior managers and directors who,

- Understand both their obligations to provide proper oversight, as well as the value that this can contribute towards their wider decision-making processes.
- Can proactively define corporate level oversight and governance systems that are appropriate for the size and nature of the business. A small family-owned business still needs corporate governance, but certainly not at the same level and intensity as

that of a major PLC – a further example of the need to ask the right questions. *'What exactly is it that I (or we) are trying to govern, and why?'*

- Commit to implementing oversight in the most efficient manner, communicating its necessity and value within all levels of the organisation, and conducting regular open reviews towards ensuring that it does not become a 'form filling' exercise. (Where appropriate, this may include proactively sharing some governance documentation at various management levels within the company, to foster education, understanding and improvement.)

7.6 Effective Communication and Delegation

US Presidential speechwriter James C. Humes once stated that, *'the art of communication is the language of leadership'*. This phrase is memorable for two reasons. First, because of his observation that leadership actually has a language, and by implication, that you cannot therefore be an effective leader unless you can speak the language. Second, because his description of communication as an 'art', and in the subsequent expansion of this theme that art takes on many different forms and that there is rarely any clear distinction in art between what is good and bad. (It is, however, always possible to understand very clearly which particular type of art (i.e. communication) – is most effective in attracting our attention and response).

As managers rise through the corporate hierarchy, there appears to be an associated assumption that an individual's communication skills will develop and increase in tandem. This is often not the case. Let's face it – some people are naturally poor communicators, and will always be so, no matter how much training is absorbed. In itself, this may be no disaster – particularly if the individual can recognise this weakness and take proactive measures to compensate as part of a senior management or Board team. This further emphasises the point made earlier about the importance of having a balanced team selection, with a blend of different skills, talents and experiences (strong communication being just one of these).

For the senior management or Board, there is little point in making all the right decisions if it is then unable to effectively inform, direct (and inspire) using these decisions with clear messaging.

Good communication is not just an outward art form – it is always good to remember that we were born with two ears and just one mouth! Strong managers and leaders will invariably also be good listeners, focused as much on the absorption and analysis of incoming data as they are in what they want to say.

An effective multi-directional communication system is also the basis for effective delegation. Senior management and Boards cannot stop merely at making decisions. These are only effective when they can delegate implementation, and where they are then proactive in following-up on the delivery of their key decisions in a rigorous manner.

Over the years, wider observations of senior management and Boards (across all industries, not just in construction) have been that the largest single factor in any disjointed or failing business has always revolved around poor or inappropriate communication and/or delegation skills in some form.

7.7 Summary

It is evident that risk management and human factors play a key role at many levels within an organisation. Nowhere is this more so than at the level of senior management and the Board. We consider that the following key attributes represent a core skill set for the effective control and management of an organisation at senior level.

- Good team composition – being a balanced mixture of different skills, experiences and abilities.
- An individual and collective appreciation of the wider corporate picture at a macro-level, including commitment to the need for, and value of, appropriate governance tools.
- Discipline and rigour in addressing strategic issues – including markets and project selection, but also in the strategic development and wider education of future senior leadership options.
- Independence of individual thought, and also the ability (and desire) to listen to and rationalise the views of others.
- A collective ability to identify the key issues and risks that matter, and to do so in a dispassionate manner that allows rational and objective decision making.
- Strong communication skills – being effective at dissipating information and decisions through the organisation and (if appropriate) also externally.
- Attention to detail in following the identified critical issues during the implementation process of key decisions made.

Of course, the above list is by no means exhaustive. It is precisely because there is no simple management tool that always works (and which individuals will consistently implement) that we will always have the need for individual thought, analysis, creativity and collaboration – this being most effectively realised within a senior management or Board structure conforming to the above core principles.

8

Structuring Successful Projects

8.1 Introduction

Projects go through several stages during their life, from concept to practical completion and handover for operations. There are many activities to be undertaken in order to create a successful project and it is essential that these success factors are properly planned and detailed and then efficiently implemented throughout the process, with constant reviews along the way.

8.2 So What Happens on Successful Projects? What Are the Key Factors that Create Success?

Before looking at this, we need to first define 'success'. We should aim to achieve the following:

- Build a quality project that meets the aesthetic, technical and functional requirements of the client.
- Delivery on time and on budget.
- Acceptable operating margins for all stakeholders. (Negotiating these requires a reasonable approach by both the client and the contractor, as covered in other chapters of this book and which can be quite difficult at times.)
- Good communications, collaboration and respectful relationships.
- Issues and claims resolved by sensible negotiation; no formal disputes.
- Job satisfaction and adequate remuneration for all stakeholder employees.
- Strong team spirit amongst the stakeholders, with a common objective to deliver a successful project.
- Smooth facility operations after completion of construction.
- Clients and end-users who are very pleased with their new facilities in all respects.

The following elements for success are fundamental, but it should be remembered that they all have potential for risk; this must be kept in mind and they should be reviewed on a constant basis.

- Competent leadership and professional teams for the client and contractor.
- Clear project objectives that are realistic and achievable – including design, budgets, and programmes.

Global Construction Success, First Edition. Charles O'Neil.
© 2019 John Wiley & Sons Ltd. Published 2019 by John Wiley & Sons Ltd.

- Professional consultants, and efficient subcontractors and suppliers that deliver quality work and products on time.
- Efficient contract management platforms and processes – it is important to document everything diligently so that the project history is retained and easily transferable when staff leave. Issues commonly arise when this is not done and staff turnover occurs.
- Careful planning and programming. Do not rush this, because the end result is invariably a direct reflection of the effort put into the planning (see Chapter 24 on the significance of the S-Curve). So-called *'fast tracking'* and/or making an early start on site to impress the client rarely works, and generally ends up costing more and taking longer.
- Realistic and achievable cost plan; contract pricing with a realistic build-up (do not 'buy' work, because if you do then Murphy's Law is bound to apply).
- Appropriately detailed specifications and contract documents (spend extra time on this – it is really worth it).
- Design development running in close parallel with cost planning.
- Healthy project cash flows, with head contractor and supply chain payments in line with programme milestones.
- All parties having a detailed knowledge of their contract documents and contractual obligations (Public Private Partnerships (PPPs) in particular).
- Robust risk management throughout the life of the project, with the key methodology being early warning through constant monitoring and anticipation, together with an inclusive stakeholder communication process.

You will note that all of the above areas have a strong human element, and it is almost taken for granted that the teams involved should be right on top of the technical aspects.

If any of these areas are badly structured or handled, then the effect on the overall project can be serious.

8.3   The Different Activities and Responsibilities, from Concept to Completion of Construction

8.3.1   Building Effective Teams

This is essential for both corporate management and for successful project delivery, and the CEO should be just as much a team member as the site manager. The human behavioural element of all personnel should be a prime consideration for the CEO and senior management when assembling effective teams.

In Chapter 10, Tony Llewellyn outlines the challenge of building teams and puts forward that there are three essential elements required to make a project work, these being technical intelligence, commercial intelligence, and social intelligence. The industry puts a lot of training and value on the first two, and largely ignores the third. The return from these values and the hard work of diligent and efficient team members is seriously diminished if the corporate management is not competent, sometimes with disastrous results for supply chain stakeholders as well as for the projects, and not least the people on site running the projects.

Most people in construction and associated industries devote their lives to it and after even 5 years, let alone 20 or more, it can be devastating to see your company collapsing around you because an incompetent CEO and dysfunctional Board have got it all wrong at the strategic and financial level, with flow-on effects on projects. At worst your pension might go down the drain, even though the incompetents at the top still take their million dollar salaries and bonuses; rewards for failure are all too common.

This is not being too dramatic. Unfortunately it is just summarising what happens all too often.

However there is a lot to be positive about, as there is clearly a new push by progressive construction companies to greatly improve the quality of their business management, their risk management, and the competency of their site management in order to increase overall efficiency, obtain the budgeted margins, and reduce the incidence of cost and programme over-runs.

This is taking place through placing more emphasis on having suitably qualified people in the team, all the way up from the site manager to the Board, and through everyone understanding the benefits of better collaboration and communication. Hopefully, the days of employing the *'dominant yelling sledge hammer'* manager that the industry has been renowned for are nearly over.

All of the above issues are very important considerations in building effective teams. In Chapter 11 we discuss these human resources aspects in greater detail.

8.3.2   Understanding the Bigger Picture

It is important that managers at all levels understand the bigger picture of how their projects are structured so that they are aware of the danger areas. There is nothing more frustrating for a conscientious manager on site or in the head office than to be treated like a 'mushroom' and not know the bigger picture. This indicates poor senior management, who cannot work out the difference between essential general information for their staff and confidential higher level information.

8.3.3   Know and Manage the Contract Diligently

It is fundamental that everyone who has management responsibility should have a basic knowledge of how contracts are structured and written, fully understand their terms and conditions in respect of the obligations and liabilities of the parties, and how to apply the contract terms to project management. How this is done will have a huge impact on the progress and efficient delivery of a construction project.

Project managers cannot implement the contract efficiently if they do not fully understand *all* the terms, conditions and specifications and the rights and obligations flowing from them. They need to know and understand the terms of contract so well that they know immediately where to look in any given situation.

It is then equally important that they carefully comply with the documentation requirements of the contract, as distinct from the technical delivery of the specifications. They must carefully comply with the processes related to administration, such as with the content and time limits for notices, variations and extensions of time. Failure to do so can be a very expensive oversight.

Unfortunately it is quite common to find project managers that do not understand how their contracts work in detail in respect of both the rights and obligations and also the administrative processes, and who rarely look at contracts until they are in trouble. For the majority of the time this trouble might well have been avoided or mitigated if they were on top of their contract and managing it efficiently. It is senior management's responsibility to ensure that their project managers are properly briefed and have the experience and capacity to manage the project efficiently.

Project Appointments

It is short-sighted and false economy to employ someone at a rate purely to meet a budget, even though it is known that the person does not have the necessary skills and competence, rather than employ someone who can comfortably do the job at the going market rate. The budget needs to be realistic in the first place.

Example: the contract for the design and construction is worth $100 m, so any variations, additional costs for extensions of time, liquidated damages etc. are likely to be significant sums, yet some out-of-touch director in the head office has absolutely refused to increase the budget by another $20 000 in order to pay for a competent person to run the project. This is very short-sighted and unrealistic. There is an excellent truism that says it is cheaper in the long run to employ two really capable people at higher rates than four much cheaper people who do not have the skills and experience.

In regard to writing the terms of contract, it is amazing how people can take a relatively straight-forward contract document and turn it into one that contains terms that are ambiguous and difficult to understand, and has grey areas or gaping black holes in respect of the rights and obligations of the parties. It is important to minimise this area of risk when drafting contracts and during the subsequent due diligence prior to signing. The parties to the contract normally work closely in conjunction with their lawyers during the drafting phase.

(Refer to Chapter 12.)

During the drafting, due diligence and checking stages of a contract it is essential to have reviews carried out by experienced 'seasoned' project managers to test the contract with some difficult hypothetical situations.

It is really helpful to project managers and others administering the contract, such as the quantity surveyors, if the terms of contract are written as much as possible in layman's language and not in convoluted *'lawyer-speak'*. Fortunately there has been considerable improvement in this direction in recent years. Likewise, there has been considerable improvement in the general format of contracts.

It is also important to not have too many cross-references in the terms, both within the main contract and between related contracts, as with PPPs where you have a suite of contracts, such as the Project Agreement, the Finance Agreement, the Construction Contract and the Facilities Management Contract. Some contracts have so many cross-references that you have to flowchart your way through the documents to find out whom is responsible for what in respect of the rights and obligations of the parties. A simple clause in the Construction Contract may just refer to the Project Agreement, but that particular term may have serious ramifications for the contractor in regard to pass-through obligations to provide facilities and be responsible for the costs thereof. The facilities manager and their obligations down the track could be affected in the same way.

Contracts that are well drafted will be beneficial for both parties and provide the framework for smooth project delivery, with minimal extra costs and disputed issues, subject to each party managing the contract efficiently and complying with their obligations.

> The author worked as project director on a multi-storey contract in Vietnam in 1993–1996, one of the first new developments in the country at that time when Vietnam was opening its doors, and we spent seven months fine-tuning the specifications and the contract in both Vietnamese and English. The parties were a Vietnamese/Singaporean joint venture client, a German contractor, Australian design consultants and French bankers. The time spent was really worthwhile because the result was a smooth, trouble-free construction that finished 42 days ahead of the contract date, with minimal variations. All parties ended up happy, with excellent commercial relationships.

8.3.4 Performance Bonds, Payment Terms, Retentions, and Pricing of Variations and Back Charges

Far too many contractors, subcontractors and suppliers enter into contracts that contain unreasonable and unacceptable terms and conditions and then get badly exploited by their clients.

There must be a level playing field. Clients obviously want their project delivered successfully in all respects, but success is not only dependent on the performance of the head contractor, subcontractors and the suppliers. The client also has to perform efficiently and responsibly. This means being willing to sign a contract that has fair and reasonable specific terms of contract, including the specifications, programme, financial requirements (performance bonds, etc.), payment terms, dispute processes, etc., and the general conditions of contract.

The author was managing director of a tier-two construction company with a steel fabrication division that operated as both head contractor and as a subcontractor to tier-one contractors, so we had two types of clients. From my experience there are a few basic rules to stick to, but the basic principle in regard to your clients is that '*we are builders, not your bankers*'.

We used the following terms as the basis for contract negotiations after we had been bitten badly a few times by unscrupulous clients and tier-one head contractors.

Each client and contract is different and the terms you negotiate will vary accordingly. Our best clients and head contractors had no problem with fair and reasonable contract finance and payment terms because what they really wanted was performance and we were first-class performers in meeting programmes and delivering quality.

If we were concerned about a client then we insisted on most of the following terms. If they objected, then most likely it was an indication of their intentions towards payments and/or a bullying attitude, in which case we did not want their business anyway. We were also fortunate that we had plenty of work available.

1) The terms by which performance bonds can be redeemed are extremely important and must be carefully drafted in quite specific terms in respect of how they can be called upon.

2) There should be an upfront establishment payment on signing of the contract to cover advance purchase of materials and site establishment. Depending on the type of contract and the work involved, this can be in the range of 10–20%.

3) Progress payments should be geared to milestones, signed off by an <u>independent</u> project Engineer, whose appointment is agreed to by both parties at the outset. We always asked for, and mostly agreed on, payment 14 days after certification by the engineer, but in any case payments should not exceed 30 days after certification, which is normally accepted by government departments and credible clients.

4) Retentions should go into an independently managed trust fund, with clear terms and conditions for their release, to be signed off by the independent Engineer;

5) Variations – the contract should contain a methodology whereby potential variations must be priced in advance by the contractor and be approved in writing in advance by the client before any work commences. The contract should include a schedule of rates for valuing the variations, or if not applicable then a reference to the use of the going market rate. This protects both parties. In case of disputes the contract should nominate an independent third party empowered to make a decision on the values.

6) Back charges by the client – the contract should spell out the methodology and means of valuation in a similar way as with the variations.

7) Delay and disruption costs for delays caused by the client – the contract should nominate the methodology for dealing with these, including a schedule of holding costs and *'Preliminaries'* to be applied.

8) In the event of progress payments not being made on time, the contract should stipulate that the contractor has the right to stop work seven (7) days after giving notice in writing, with no penalties applicable and with nominated holding costs payable (as in point (7) above) until work restarts or is formally terminated by the client. *Note that we never had to invoke this clause.*

9) Liquidated damages – most contracts included provisions for LDs, but we never incurred any.

10) Penalty/bonus, in lieu of LD's. This is not a common arrangement, but we really liked it. The general rule is that you cannot have one without the other, so typically you might enter into a contract that has a fixed completion date, subject to any agreed EOT's under the terms of the contract, and fix the penalty/bonus at say $3000 per day either way. The largest that we had was US$10000 per day;

11) Final payment – if you are really concerned about the financial standing and intentions of the client/head contractor you could ask for the final 10% to be lodged as a bond with a third party, e.g. lawyers or bank, to be released on final certification by the independent Engineer. We only ever asked for this with new clients that we were unsure about.

In summary, the main objective is to structure your cash flow so that it is always positive in terms of receiving more than your own costs to date and what you owe creditors, at every stage in the programme, so that if the project stops for some reason or other, including the client going bust, then you will not be 'out of pocket'. Realistically, this can be difficult to achieve on some contracts, but you should only allow yourself to get into a negative cash flow position if you have absolute faith in the client's solid financial situation and integrity.

> **A perfect project is when people say**
> *'Weren't they lucky to have such a smooth run – what did they do to earn their money'*
> **But it has nothing to do with luck!**

8.4 Checklist for Structuring Successful Projects

The following processes and procedures can be used as a checklist for most construction projects. They will nearly all apply to PPPs, but to a much lesser extent with D&C or construction-only projects.

The parties responsible for the following processes are identified in italics *(client, contractor, user/operator, facilities manager (FM)):*

8.4.1 Structuring the Project (*Client*)

- Concept – the schematic design and the business case
- Client financing – concept stage and whole of project
- Site acquisition
- Planning and other authority approvals
- Feasibility studies
- Preparation for tender – assessing the overall project; the common sense of the concept
- Assessing the degree of readiness to proceed
- The client's own resources and competency
- The political situation
- Calling for expressions of interest (EOI's) and planning the tender.

8.4.2 Tendering and Bidding Activities (*Client, Contractor, User/Operator, FM*)

With traditional contracts it is essentially the responsibility of the client to manage the process, but with D&C and PPP projects there is also substantial involvement from the contractor, the future operator (end-users, e.g. National Health Service) and the facilities manager.

- Preparing the tender documents – drawings, specifications, special terms and conditions
- Design development and cost planning in parallel for D&C and PPP projects
- Establishing the risk management processes
- Consideration of operational items – end-user requirements, O&M, life cycle replacement
- D&C planning and programming
- Firming up prices – subcontractors and suppliers
- Bid qualifications
- Due diligence on contract documents
- Client check of financial strength and experience of all participants in the Bid (should have occurred during assessment of EOI).

8.4.3 Establishing the Risk Management Process (*Each Party is Responsible for their own Risk Management*)

The extent to which this is done will depend on the size of the project, but at the very least a risk management schedule should be prepared that covers all areas of the project. Large projects should appoint a specific Risk Manager whose task it will be to consolidate and monitor all of the potential risks that are nominated up-front by the different section managers. Refer to Chapter 17 for further elaboration on this point and a typical team structure and flowchart for controlling risks.

8.4.4 Finalising the Financing and the Contract

- Establish realistic and carefully prepared budgets and cash flow projections for progress claims and milestone payments *(client, contractor)*.
- Have a realistic contract sum analysis and scheduled rates for overheads, preliminaries and margins *(contractor)*.
- Agree fixed margins in the contract for modifications/variations for the contractor, facilities manager and equity partner in the case of PPPs, so that arguments over future variations are minimised *(client, contractor)*.
- Ensure the financial structure has some flexibility and headroom for variations *(client)*.
- When preparing PPP contracts, place heavy concentration on the services scope and obligations in the contracts *(client)*.
- Test the contract as much as possible to eliminate black holes, grey areas and ambiguities *(client, contractor)*.

8.4.5 Establishing the Project Leadership and Team Spirit

Early appointment of a top-line project director by *each of the main stakeholders*.

Establish and demonstrate real leadership from the outset, through a stakeholder workshop held immediately after signing the contract, or financial close for PPPs *(client)*:

- To establish relationships, relationship management and guiding behaviours
- Commence building trust and respect
- Understand the needs and success factors of all stakeholders
- Establish common goals; agree key objectives and mission statements; e.g. *'a world model PPP' and 'no disputes'*
- Generate positive attitudes and create team spirit.

Run Review Workshops at least every four to six months – the main things that will need 'renewal' will be relationships, communications protocols, and project objectives – remember that staff changes are always happening *(client)*.

8.4.6 Agree the Key Processes with *All Stakeholders* at the Outset

- Control and reporting requirements and procedures – client, project or SPC manager, contractor, design consultants, FM.

- Management and document control platforms with different levels of access, e.g. Affinitext, Team Binder, Incite, SharePoint, etc.
- Communications protocols – establish clear rules – no side-plays.
- Meetings and working groups – strict rules – with timetables, mandatory attendances, agendas, delivery times for minutes – chaired and proactively led by project directors.
- HS&E regimes.
- Crisis management planning (see more detail at the end of this chapter).

8.4.7   Construction Phase (*Contractor, FM*)

- Financing, staffing and allocation of resources
- Org charts and job descriptions
- Planning and programming
- Detailed design development, again in parallel with the cost plan
- Full involvement of the FM from day-1 – 'whole of life'; O&M practicality and optimisation; 'granite can be cheaper than bricks over 25 years'.

8.4.8   Commissioning, Completion and Transition to Operations (*Client, Contractor, User/Operator, FM*)

- Ensure enough time is allowed for commissioning and provide adequate resources – PPPs can take far more time and resources than normal D&C projects, which traditionally might only be 95% finished at practical completion and the remainder is completed during the defects liability period (see Chapter 15 on PPPs).
- It is a common error to cut short on resources and the time allowed when budgeting, but this can prove expensive in the long run.
- Programme the activities carefully, with participation and agreement from all stakeholders, particularly the client, user/operator and the facilities manager, who will take over full control during the transition period.
- Do not confuse completion, commissioning, and transition. Treat them as entirely separate activities and plan, programme and resource them accordingly
- Experienced commissioning and transition managers should be engaged – the transition manager's role is to liaise with and coordinate the activities of all stakeholders and is not the same as the commissioning manager, who comes from the contractor

8.4.9   Defects Liability Period (*Client, Contractor, FM*)

- Plan and programme the management and resourcing for the rectification of the defects in conjunction with the relevant subcontractors.
- PPP projects, hospitals and prisons in particular, are more difficult than traditional construction or D&C projects as it is necessary to be 99.9% defect free at project completion, for two reasons:
  - The independent tester will require it under the contract
  - Access for fixing defects will be restricted as soon as the facility goes into the operations phase.

- Form subcontractor *'task teams'* for an efficient rectification process, comprising a skilled tradesperson from each specialist subcontractor involved with defect rectification (snagging in the UK), probably four or five of them in total, and have them work together as a team, e.g. in a hospital, room by room down a corridor, securing each as they are signed off defect free. *Note: this is the most practical, efficient and cheapest method.*
- An interface agreement should be in place between the contractor and the facilities manager, with the FM being responsible for all but major items involving warranties, such as large equipment breakdown.
- It can also be cost effective for the contractor to subcontract the management and completion of the defects to the FM after the bulk of them have been fixed.

8.4.10   Claims and Disputes (*All Parties*)

- Establish clear methodologies in the contract and in the early stakeholder working sessions for reviewing and processing claims. Stick to strict time schedules for processing claims, because they only get harder the longer they are left.
- Ensure that the contract includes a workable dispute resolution section. All too often this is left to the last minute and something is included in a rush without proper thought.
- In structuring the dispute resolution process, ensure that there is an emphasis on negotiated settlements, with plenty of opportunity for this to take place.
- In D&C and PPP projects, encourage the parties to adopt value engineering 'trade-offs' when design and equipment changes take place during design development, with the objective being a cost neutral result. This is a good way of minimising potential future disputes over variations.

8.4.11   Crisis Management Planning (*All Parties*)

A crisis on a construction site is very much a human behavioural situation and you can never be sure how people will react under the stress of a crisis. It is important that each project establishes procedures at the outset for crisis management and crisis communications and that everyone working on the site or connected with the project through one of the stakeholders is inducted into the procedures.

The two crisis procedures, management and communications, are equally important; e.g. if there is a serious accident on a site the treatment and evacuation of the injured person has paramount importance, but at the same time it is essential to communicate details of the incident promptly and in the right manner to the family of the injured person and to the CEOs of the construction company, the client and the subcontractors that are involved, because the last thing you want is for them to hear about it first from a reporter or on the TV news.

There are a number of potential crises that can occur on a project and procedures should be developed to handle these different areas of risk.

The **'Project Crisis – Internal Rapid Escalation Plan'** (see Figure 8.1) is a typical crisis communication plan for stakeholders, which can be extended to cover the media and the community.

PROJECT CRISIS - INTERNAL RAPID ESCALATION PLAN

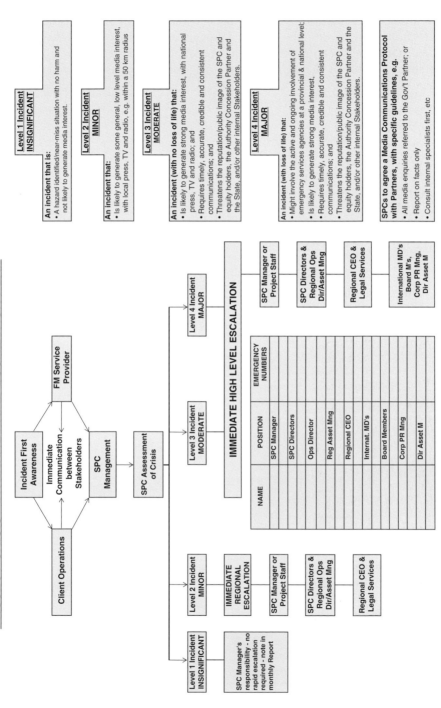

Figure 8.1

8.5 Summary

There is absolutely no substitute for thoroughly detailed planning. Successful projects don't just *'happen'*. They directly reflect the thought and degree of detail that goes into the planning and preparation of the documentation for all stages, followed by efficient implementation. Only by doing this will you achieve *'success'* as we defined it at the beginning of this chapter.

9

Understanding and Managing Difficult Client/Contractor Relationships

David Somerset

Managing Director Somerset Consult UK

9.1   Introduction

Construction projects are invariably high-risk ventures and, as with all projects, their success depends heavily on the successful management of risks. The collective responsibility for management of risk lays with the project stakeholders themselves and so the relationships which develop between those persons and organisations are fundamentally important.

What do we mean by 'relationships'? A dictionary definition might be *'the way in which two or more people or groups regard and behave towards each other'*.

In a construction project relationships may be additionally defined as 'contractual' (the parties to the contract) or 'professional' (consultants) and it should always be recognised that a 'personal' element frequently develops.

At the outset of most projects enthusiasm and expectation levels are high and relationships blossom on a wave of optimism. However, there is a high risk[1] that relationships with clients and contractors will deteriorate to some degree as the project progresses. A well-known case, *Walter Lilly v Mr McKay*,[2] highlights the extent to which relationships can deteriorate, ultimately giving rise to legal disputes.

For the purpose of this paper, rather than discuss generic issues of client relationships, I will focus on a specific type of client – one for whom relationship issues are perhaps familiar territory. As for contractor relationships I will focus on those issues that arise after the client/contractor relationship has broken down. However, in both cases it should be noted that much of what follows has a wider relevance and application.

9.2   Problems Posed by Difficult Clients

In my experience wealthy residential clients are, by and large, the most challenging, and their projects tend to suffer higher than average levels of delays, cost over-runs and defects.

---

1 This risk is much reduced when client and contractor have previously worked together.
2 Walter Lilly & Company Limited v Giles Patrick Cyril McKay and DMW Developments Limited (2012).

Typical characteristics of this type of client include:

- They are used to getting what they want without much resistance.
- In the case of foreign investors; language, cultural and technical obstacles are common. For example, the difficulty of understanding planning restrictions.
- They will often make changes to the scope of works regardless of the progress of the work on site and the resulting impact.
- Sometimes one is dealing with a husband and wife team, neither of whom seems to communicate with the other.
- They are impatient and do not understand time scales required for various stages of the construction process, for example to prepare design and tender documents.
- They do not wish to understand the concept of claims, for example extension of times/monetary claims. This they interpret as some form of wrongdoing on their part.
- Their involvement in a project may be intermittent, they flit in and out.
- These clients are often represented by persons close to them, acting as agent and reporting back to the client. This can work effectively but can also prove disastrous if that person seeks to filter important information that they know will not be well received by the client; for example, increased costs or delays. When ultimately the client does become aware, there are often serious consequences.
- They do not understand their obligations or the need to make regular payments.
- They are often under the illusion that they are paying for a finished product irrespective of how that product was procured.

9.3 How to Manage Difficult Clients

It might be argued that any client is potentially a difficult one but recognising the characteristics described above should sound alarm bells.

There is no overarching solution to managing difficult clients but there are obvious areas where mistakes can be made and from which problems commonly arise.

9.3.1 The Client Brief

The responsibility for agreeing and recording the client brief (and later revisions to it) generally falls to a project manager or architect (or both). If the client brief is insufficiently detailed then problems are certain to follow. In the case of wealthy clients it can be difficult to gain sufficient access to them for this exercise. Moreover, wealthy individuals are commonly more used to 'hand waving' than being pinned down for more exact information. There can also be a tendency for consultants to think that a wealthy client will be more tolerant of additional costs arising from late design development. The truth is that wealthy clients are often unsure of exactly what they do want but are very sure of what they don't want when they see it.

So, the first step in managing a difficult client is to properly define the client brief, including not only specific requirements but also the more general (as yet undefined) requirements and how future design and procurement processes will apply to them.

9.3.2 The Programme

Difficult clients are generally impatient and often pressurise their consultants into accepting unrealistic timescales for design and on-site construction of the works. For example, on a project for a fit-out for a high end residential apartment, the interior designers (a well-known practice) advised the client that the works (valued at circa £7 milllion) would take eight months from placing orders to final completion. This included manufacture of circa £2 million of bespoke joinery. Given the quality of the finishes expected by the client, the duration was tight with no float in the programme. The client however insisted that the project be completed in four months and when the interior designers explained this was not possible, their appointment was terminated immediately and without further discussion. A 'one-man band' consultant was approached and, not surprisingly, appointed to manage and procure a contract to complete the works in four months. Needless to say, the works were not completed in 4 months, nor in 8 months, but in 12 months! The consultant and contractor are now in dispute with the client over claims and counter-claims. The lesson to be learnt is that whatever resources might 'thrown' at a project, in particular those which include high quality finishes, there is always a minimum construction time.

My recommendation to any consultant or contractor faced with a client expecting an unrealistic period for the works would be to try and explain why a longer time period is necessary, failing which the sensible option would be to withdraw. Although this course of action may be unpalatable it is certainly preferable to the prospect of a future legal dispute.

9.3.3 Cost of the Works

The potential problem of clients with unrealistic programme expectations not only affects design and construction but can impact on cost control as well.

The task of a cost consultant (appointed by the client) in estimating/cost planning and cost control for a high-end residential property can be challenging unless the design has been substantially completed. In my experience, this is very seldom the case with the specialist fit-out elements of the works often being included in contracts as provisional sums. On one fit-out contract where the client insisted that the works commence whilst the design was still in the concept stage, the cost consultant, not willing to stand up to or challenge the client, duly prepared a tender document. The lowest tender Bid was £12 million but this comprised £10 million of provisional sums. As the works progressed the client constantly changed his mind and massive variation costs led to the final account for the fit-out works being agreed at £22 million. The cost consultant's role on the project became no more than reporting the costs. Once the project was completed, the client alleged the cost consultant had failed in their duty and a legal dispute followed – which still remains unresolved.

With regards to costing variations and/or changes to the scope of the works, it is significant that most fit-out works are carried out by specialists who tend not to provide detailed cost build-ups but rather lump sums. Unless the cost consultant has obtained cost breakdowns or rates prior to agreements being executed, post-contract negotiations with these specialists can be difficult and they lead to protracted arguments. A cost consultant engaged for this type of project needs to be experienced enough to avoid being held to ransom by the specialist.

9.3.4 Quality Procedures

Quality control procedures are an important element in all projects. However, with high-end residential projects, the need to implement effective quality control/defect remedy procedures is even more important. The finishes on high-end residential projects are:

- More expensive than 'standard' finishes.
- Applied or installed by craftsmen who take pride in their workmanship and tend not to respond favourably to working to tight deadlines.
- Often subject to long lead-in times. For example, on one project, the stone for a bath was imported from Brazil, but having been shipped it was discovered there was a flaw in the material. The stone had to be re-ordered; this involved it being mined, dressed and then shipped. The delay to the works was over four months.
- Tolerances with some of the finishes can be more exacting.

The need for quality control and defect remedial works go hand-in-hand. The importance of these needs to be explained to the client as this will also help manage his expectations. On one particular project, the client insisted, against all advice, on achieving completion of the project a week before the 2012 Olympics in London. The project was completed but with the quality of finishes being compromised. There then followed a protracted period of defect rectification culminating in a dispute with the contactor.

9.3.5 Relationships with Consultants

Many consultants would regard an appointment by a wealthy client to be prestigious. In my experience, consultants should only accept these appointments with their eyes open and a realisation that such clients are often:

- Extremely demanding
- Prone to changing their minds and often with no understanding of the status of the project
- Impatient
- Fickle
- Lax over payment of fees
- Liable to terminate an appointment without much thought.

This is not to say that one should not accept this kind of appointment, but if one does then expect it to be demanding and act accordingly.

9.3.6 Claims

A high-end value client will often spend significant sums on the works and may even acknowledge the cost of variations. However, when it comes to claims there is often a lack of understanding and appreciation of their liabilities. Such clients often question why they are being asked to pay for something that is not tangible. Avoiding claims is therefore high priority with these clients.

9.3.7 Payments

Wealthy clients do not always understand the need to make regular payments; this situation being worse with clients who do not reside in the country where the project is located. In these circumstances it is advisable that a separate bank account with adequate funds is set up in the appropriate country.

9.4 Problems Posed by Difficult Contractors

Contractor organisations might develop a reputation for being difficult for various reasons including:

- Inadequate management
- Poor workmanship
- Taking an adversarial approach to contracts
- Aggressive pursuit of extras and claims.

For the purpose of this chapter, I describe below a situation where a contractor that ultimately displayed all of the traits listed above took advantage of inadequacies in the consultant's performance.

The contractor, which I will refer to as 'Cheap O', was one of six tenderers invited to submit a Bid for a new high-class residential building. The tender documentation was poorly prepared, with the scope of the works not clearly identified and with a significant value of provisional sums and elements where design responsibility was otherwise ambiguous. The tender documents required the successful tenderer to provide a performance bond.

Cheap O submitted a highly competitive tender Bid that was 10% less than the second lowest Bid and considerably less than the quantity surveyor's cost plan. Furthermore, the construction duration was 14 weeks less than the second tender and again less than the quantity surveyor's time estimate.

Each tenderer was requested to set out their Bid as a detailed contract sum analysis. Cheap O's contract sum analysis consisted of four pages of mainly one-line items. The other tenderers provided more detailed and comprehensive contract sum analyses.

The quantity surveyor, rather than undertaking a detailed and in-depth review of the tenders, recommended the appointment of Cheap O. The client accepted the recommendation, knowing the Bid to be highly competitive. Up to this point no financial check had been made on Cheap O. However, a request was made to provide their latest financial statement – this was provided, albeit that it did not reflect the current financial year. The information provided identified a significant loss in the preceding financial year. However, no further financial checks were made and a letter of intent was issued for the full contract sum. Whilst the contract documents were prepared, they were not sent to Cheap O.

Cheap O commenced work on-site, based on a 45 week construction programme.

The works commenced extremely slowly, with Cheap O issuing numerous requests for information. The contract document was sent to Cheap O for signing but was never returned. Requests were made for provision of the performance bond but they were met with excuses from Cheap O.

At week 45 (the original contract completion date), less than 40% of the works had been completed. Some weeks later, Cheap O's contract was terminated, followed by Cheap O going into administration.

The works, on the date Cheap O left site, contained a significant number of defects. No contract had been returned and there was no performance bond.

The task of appointing a replacement contractor was challenging and a contractor was appointed on a 'cost reimbursement' basis. Some two years later, the project was completed with the final account being in excess of double Cheap O's tender Bid.

9.5 Steps to Manage Difficult Contractors

There are a number of steps that can be taken to mitigate the risks associated with difficult contractors.

9.5.1 Developing a Procurement Strategy

It is essential to adopt an appropriate procurement strategy in response to the client's requirements. The RICS Guidance Note, *Tendering Strategies*[3] sets out practical advice for establishing a procurement strategy.

Every project will have a business case, with various factors such as:

- Funding
- Time
- Performance
- Capital versus operational cost
- Risk
- Type of project.

Once the procurement strategy is established the chosen procurement route can be developed – be it a lump sum contract, design and build, construction management, or cost reimbursement.

Whatever procurement route is chosen, it is of the utmost importance that the tender documents align with the procurement route. In the case study above, the tender documents were not appropriate for a lump sum contract. The tender documents should have clearly identified the full scope of works with no opportunity for the tenderers to apply their own interpretation to ambiguity/gaps in the scope of works.

9.5.2 Tender Reconciliation

Whilst the review of tenders can be undertaken on a documents only basis, in the case study, there should have at least been a meeting(s) with the two lowest tenderers to discuss and clarify the tender Bids.

9.5.3 Acceptance of the Tender

Before there is any acceptance of a tender, there should be financial checks and, where a performance bond is required, ensure a bond can actually be obtained. In the case study, limited financial checks were made and no enquiries made as to whether Cheap O could obtain a performance bond. In the event that a letter of intent is issued (which

---

3 Published by the Royal Institution of Chartered Surveyors, January 2015.

should if possible be avoided[4]) then it should have a defined financial cap. This enables the contract status to be reviewed once the cap is reached.

9.5.4 Contract and Other Documents

Ideally the contract should be issued before commencement of the works, including provision of a performance bond and warranties. Once the contractor has commenced works, the client loses leverage to obtain these documents.

9.5.5 Performance on Site

The contractor's progress should be monitored carefully. In the scenario above, even a simple review of the monthly application for payment would have revealed that the payment drawdowns were less than should be anticipated – a good indication of underlying problems. These problems could have been due to outstanding information/cash flow or resource issues. In the event that lack of progress is attributable to the contractor, then action should be taken. In the above scenario, action was in the form of low-key meetings with hollow promises made by Cheap O. Had positive action been taken this may have resulted in the earlier termination of Cheap O's contract. This would have avoided the situation of allowing Cheap O to remain on site for months with minimal progress and poor workmanship. To allow Cheap O to continue on site in the hope that progress would improve was a totally unrealistic.

Managing a difficult contractor starts with the preparation of the tender documents that are appropriate for the procurement route. Careful review and analysis of tenders is essential and enquiries as to their financial standing should be undertaken. Attempting to manage a difficult contractor from a base of poorly drafted tender documents and inadequate tender analysis is a hopeless position. Letters of intent should be avoided and an executed contract should be obtained before the works start, together with any performance bond or other security that is required. The performance on site of Cheap O should have been adequately monitored and positive action taken at the appropriate time.

9.6 Conclusion

Anyone with experience in the construction industry will be aware that both clients and contractors can be difficult to deal with. The more experienced will have learned to recognised the warning signs early enough to establish a suitable management strategy or alternatively to decline an appointment. What would a suitable management strategy involve? The answer is unsurprising – adopting a rigid 'best practice' approach for your profession. If a consultant allows a client to pressure him into compromising standards then it is he who will ultimately bear the associated risks. The more difficult a client is likely to be, the more tightly the consultant needs to adhere to professional standards.

---

4 The dangers of a letter of intent are explained in the case of The Trustees of Ampleforth Abbey Trust v Turner Townsend Project Management Limited (2012).

Similarly, difficult contractors may seek to take advantage of clients and consultants who cut corners. Opportunities to weed out potentially difficult contractors at tender stage should not be missed. Thereafter, effective controls can only be applied by a rigid application of the contract – which is only possible if the contract itself has been properly established.

10

Social Intelligence – The Critical Ingredient to Project Success

Tony Llewellyn

Collaboration Director at ResoLex Ltd, UK

10.1 Introduction

Is construction a technical or a social process? This is a question I frequently ask when lecturing on the subject of building effective teams. The instinctive reaction of the majority of any audience is a clear sense that building is a technical activity. When I then enquire about the skills need to become a successful builder or designer, the responses focus on knowledge of structures, materials, programming, and contracts. There are, however, usually a few more experienced members of the audience who express a different point of view. They have learned that construction is fundamentally about people and the extent to which they are able to connect and communicate with each other. This chapter will focus on the challenge created by a general lack of comprehension of the need for strong social skills, and suggests how the industry might adapt its practice for developing the competences and capabilities of the young, and even not so young people, who work in our industry.

The title of this book is *Global Construction Success*. The underlying theme of the different contributors is the need to move beyond the technocratic mind set prevalent in our industry and to recognise the risks posed by failing to recognise the essential role humans play in the process of designing and building structures. Developing a new perspective on the process of construction is important for many reasons, not least because so often the root cause of failure in construction projects is not the integrity of the building or structure, but the failure of humans to operate as a cohesive team. Many construction projects suffer from a pattern of behaviours based on a transactional mind set. The common assumption is that all buildings can be designed, priced, and delivered within a defined contractual agreement that requires no attention to building collaborative relationships based on trust. As the project moves outside the contractual parameters and the costs start to increase, the transactional mind set prompts a need to apportion blame, rather than seek to find a win–win solution.

It is easy to see how this limiting mind set evolves. When we first start our training, the focus in the educational system is to learn the *science* of the numerous elements that go into designing and building a structure. We are taught the technical 'competences' in sufficient detail that we can begin a career doing something that is of practical use to our employers. As careers progress it becomes necessary to understand the commercial aspects of the construction process, such as pricing, contracts, and how to manage risk. Sadly this seems to be where most of the learning and development resources applied to the construction industry stop.

10.2 Project Intelligence

Building is a human activity. Generating something new from 'thin air' requires groups of people who are able to conceive, plan and then execute a series of complicated activities that will, in some form or other, enhance the lives of other human beings. Large construction projects require the input of many highly specialised skills, which are of limited value themselves but are of significant worth when combined with the skills of others. It is this combination of applied skills that leads to success. Project teams need people intelligence or, more particularly, three distinct types of intelligence.

- *Technical intelligence.* The knowledge and awareness of how a building or structure is designed and assembled.
- *Commercial intelligence.* The knowledge and awareness of issues around money, contracts and the identification and management of risk.
- *Social intelligence.* The knowledge and awareness of how humans behave in groups and teams.

Large projects require lots of people. Despite advances in automation, we remain a long way from having robots to design, manage, and deliver complex engineering software, logistics or construction projects. These enterprises require human expertise and ingenuity to make them happen. As illustrated by the Venn diagram in Figure 10.1, there is a sweet spot where these three elements combine to form the rounded capability to lead and manage a major project. When assembling a project team, we tend to value specialist technical and commercial skills in others because they cover gaps in our own skill or knowledge set. Adding social intelligence to the group, however, tends to be seen as much less important.

Figure 10.1 The three components of project intelligence.

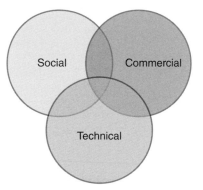

10.3 Social Intelligence

We often use the word 'dynamics' to describe how people interact when they are together in a group or team. The word dynamics actually means 'forces of change'. We tend to use the phrase human dynamics when talking about how people interrelate when they come together as a group or a team. So, human dynamics are essentially the forces that create or destroy the momentum of interaction between groups of individuals.

The problem is that people are messy. Humans are driven by a complex series of emotional reactions that are genetically hardwired into us to thrive and survive. Our behaviours are driven by motivational factors that are often not visible to others and can often be difficult for others to comprehend. To understand others and then have them understand you is an essential human skill. The ability to first understand others and then be understood by them is often referred to as *social intelligence*. This is a term used to describe a degree of awareness of the human interactions between oneself and others and also between others.

Social intelligence can be defined as *'the ability to acquire and apply knowledge and skills'* (Oxford Dictionaries, Oxford University Press). The concept has a variety of possible definitions but at its simplest, it is *'an ability to get along well with others and to get them to co-operate with you.'* (https://www.karlalbrecht.com/siprofile/siprofiletheory .htm).

Daniel Goldman (2006) observes that social intelligence has two primary components:

- *Social awareness.* The ability to sense and understanding the emotions of others. The skills include empathic accuracy, understanding other people's thoughts, feelings and intuitions, and social cognition to recognise the unspoken social norms of a group.
- *Social filters.* The ability not just to sense but also then act on the information gained. Social facility skills in teams include an ability to present oneself effectively and to influence the outcome of social interactions in the group.

Goldman argues that whilst IQ is something that you are born with, social intelligence is regarded more as a learned skill than it is a genetic trait. This leads us to the question that if social intelligence is an essential component of project success, then why do professional bodies pay so little attention to its development?

My observation is that there tends to be a tacit assumption in many teams that everyone on the team will automatically have the social skills needed to get along well with each other. This often turns out to be a significant misjudgement. Whilst humans have evolved to work in groups, left to our own devices we also have an amazing capacity to find reasons to disagree with and dislike other people or groups. As projects have got bigger and more complex, it is clear that future generations of project leaders are going to have to take the development of their social intelligence more seriously.

Positive social interaction in project teams cannot therefore be left to chance. In just the same way as for technical and commercial competencies, there are methodologies for building strong social bonds. Those methodologies need to first be understood, learned, and then practiced.

We all learn to do it instinctively as we develop from children to adults. This learning is however innate rather than conscious and so it is difficult for most people to articulate. Difficult, however is not the same as impossible. It is quite within our abilities to understand the mental and behavioural patterns of others. However, it does require some knowledge and some effort. As the field of psychology has moved from working with people who are mentally ill to those that are generally mentally well, a large body of information and knowledge has been developed on how we behave in groups. The problem for construction is that in our industry such knowledge has been historically deemed to have little direct relevance to the design and engineering professions.

10.4   Learning and Development

Social intelligence is a learned skill set. Whilst some people have a stronger sense of how to interact socially than others, everyone is able to learn the processes and routines that build communication between adults. As children we typically have few problems engaging with our peers but, as with so many aspects of natural behaviour, we often become more awkward as we grow older.

The prevailing belief in universities and colleges appears to be that the development of social skills should be left to future employers. The problem in deferring the development of these skills is that too much effort is spent on remediating bad habits and rebuilding confidence, rather than providing the more advanced social competences needed to help individuals and teams thrive in a complex environment. It is interesting to note the outcome of a survey of teachers, reported in an article in the *Times Literary Supplement* dated 30 August 2017. The article highlighted the statistic that 91% of respondents felt that schools should be doing more to help their pupils develop teamwork and communications skills.

There is a tendency to call the ability to understand and communicate a 'soft' skill, the implication being that learning how to connect with other people is an imprecise art rather than a precise process. I believe this is an outdated description emanating from a time when managers lacked both the knowledge and the need to recognise the practice and process of efficient and effective team development. The other argument for degrading social skills to the category of 'soft' is a belief that they cannot be measured and therefore cannot be managed. This again is simply lazy thinking. There is no difference in my mind in assessing someone's technical, commercial or social capabilities. It is simply a matter of choosing the metrics and then making the effort to collect the data.

Organisations will periodically invest in soft skills training in areas such as presentations, negotiation, time management, and communication.

These are active elements which reflect the 'tell' mind set predominant in western organisations. Social intelligence is a wider concept that involves more than just what one person says or does. To be able to influence what happens in a team, you need to understand what is going on. Social intelligence learning includes areas such as:

- *Systemic thinking*. Seeking to identify all of the different factors that have created a situation. It involves techniques to look at a problem or issue from a number of different perspectives. Rather than reacting to an event by jumping to an immediate conclusion based on your instinctive perception of cause-and-effect, systemic thinking encourages you to understand the influence that wider systems will have on a particular situation.
- *Understanding group dynamics*. Learning how to 'read the room' to gauge what is happening underneath the surface behaviours of the team. The skill entails recognising the subtle clues that indicate how people are reacting to each others' nonverbal communication, through body language, tone of voice, and eye movement.
- *Influential enquiry*. Controlling and steering conversations through the use of skilled questions.
- *Conflict management*. Managing difficulties and problems by creating an atmosphere that encourages dialogue rather than heated debate.

- *Psychometrics.* Understanding others by learning to recognise their motivations and preferences and adapting your messages so they recognise your point of view more quickly.
- *Sensory awareness.* Knowing how to really listen to what others are communicating, not just in words, but also in posture, eye movement, facial expression, and pace of voice.
- *Reflection.* The habit of methodically taking time to think about recent experiences and consider what happened, why it happened, and what to do the next time a similar situation occurs.

The skills set out above can be taught but as with any skill must be practiced to become proficient. These skills must be learned and developed in the same ways one acquires technical and commercial skills. The initial information can be acquired in the classroom, or from books. The real learning only comes when trying to apply this knowledge in a real life situation.

10.5 Building Cohesive Teams

Projects create a network of interdependencies. Each role relies on several other people to have completed a range of tasks to a certain standard and within a given time frame. When the tasks are clearly understood, the project has the right resources, and the time scales are flexible, then most people can manage to work together comfortably. What happens when conditions are not so benign? One of the predominant features of working life in the twenty-first century is the high level of uncertainty in the external environment. Technical change and political upheaval create a range of dilemmas for leaders and managers as they try to anticipate the future and react accordingly.

The extent to which a complex project succeeds or fails can therefore be critically dependent on the extent to which the humans involved in the project build the necessary relationships. Few people would argue that teamwork is unnecessary. The reality however is that little resource is allocated to establishing and developing effective collaborative teams. I regularly hear a variety of excuses for this low level of investment, but the most common reason is an underlying belief that provided everyone uses their 'common sense' then social connection and collaborative interaction will happen naturally.

There is too much reliance on a one-off 'team building' event after which everyone should be able to 'get along fine'. The problem is that being friendly is not the same as working effectively. A group of people may find they have a lot in common and meetings are light hearted and enjoyable. That doesn't mean they will now be more able to manage the difficulties of working to tight deadlines and conflicting priorities.

This naïve approach to team development would be understandable if there was no information available upon which to build a knowledge base. There is however a substantial volume of research that has gone into trying to understand the common factors that will take a group of people and turn them into a cohesive unit.

10.6 Introducing a Specialist into Your Team

Training project managers and team leaders to develop their social intelligence skills is not a radical proposition. A more stretching concept is to introduce a specialist in team dynamics into a construction team. Bringing another consultant to sit around an already crowded meeting table might initially seem to add to the 'noise', adding further layers of administration and cost. Project managers already complain that clients are often unwilling to pay them to put in the time required to deliver the quality of service that they believe a project needs. The challenge of educating clients to understand value is a topic for another chapter. Here, I want to make the case that adding specialist social skills and capabilities into a project team is likely to demonstrate sufficient value that the costs are irrelevant.

The key to success for most clients is that the project is delivered on time. My research into successful and unsuccessful projects shows that one of the features of having strong social skills in a team is that projects have an increased chance of staying on programme. This is not as tenuous link. Teams who are able to work as a collaborative group tend to engage in dialogue over problems and are able to resolve issues much more quickly than those whose instinct is to resort to contractual debate.

The first part of this chapter has focused on the need for improved social intelligence in a project team. If the skill set is not already present in the key members of the project leadership then it is logical that one should find someone with those skills to add to the team's collective abilities. The proposition is not as radical as it sounds. There are, for example, a number of contracts frequently used in the United Kingdom that require a nominated individual to carry out a role designed to smooth the relationships between the key parties to the contract. The NEC4 Alliancing contract proposes the use of an 'alliance manager' whilst the FACI and TAC-1 contracts published by the Association of Consultant Architects provide for the use of a 'partnering adviser'. These roles are intended to keep the team focused on maintaining the spirit of the legal agreements to the benefit of all of the parties involved.

10.7   Coaching the Team

A 2017 report by Pinsent Masons titled *Collaboration Construction 2: Now or Never?* considers the importance of adopting a new set of collaborative working practices to increase efficiency, reduce cost, and improve customer satisfaction. One of the many recommendations of the report is to make use of a team coach. The authors of the report describe the team coach as having an independent role, where the focus is on helping the team achieve the desired project outcomes rather than working with individuals.

The concept of team coaching is rapidly gaining acceptance in many industries and organisations around the world. Team coaching in sporting activities is a well established and recognised role. Many sporting bodies have established codified coaching processes that have then been extended into coaching qualifications. In the business world, the team coaching methodologies and processes are still being developed but many of the executive coaching associations have begun to work out how to structure team coaching, training, and assess competence.

10.8   Managing Behavioural Risk

This chapter has focused on the need for project leaders to pay attention to the social or people based factors in addition to the technical and commercial challenges that every project must deal with. Successful projects almost always feature successful teams where collaborative behavioural norms had been embedded early in the project life cycle. Ineffective teams will always struggle to deliver on their promises and so anyone studying the risk of default must pay attention to the dynamics present within the core team.

In Chapter 19, my codirector Ed Moore and I set out some ideas to set in place the early warning signals of potential team dysfunctions. To close this chapter, however, I would make a final point around the need to expand the team's interpersonal skill set. As mentioned above, the construction industry can repeatedly point to projects where the root cause of under performance can be pinned on a failure of humans to match the behavioural expectations of the team members when the project started.

If you want to ensure such behavioural issues do not derail your next project, will you focus your attention on the technical, the commercial or the social risk?

11

Practical Human Resources Considerations

11.1   The Changing Job Requirements in the Construction Industry – Government and Corporate

The roles of construction company managers and government infrastructure managers have been undergoing some very necessary changes in recent times in respect of their qualifications and responsibilities. These have been driven by four key requirements:

- Department heads, CEO's, CFO's, and COO's need *a harder-nosed business approach*. This demands much stronger financial and corporate management skills; a disciplined approach to regulatory compliance; and common sense strategies that will minimise the risk of budget blow-outs that cause project and corporate losses.
- Project and construction directors/managers need a *broader range of skills*, so that they really understand the roles and skill sets of everyone in their team, including the IT specialists (see next bullet point). It is no longer sufficient to say that being a good M&E engineer or quantity surveyor (QS) qualifies you to become a project manager, with the idea that having got the job you can then learn on the run (refer to the fuller explanation in Section 11.2). Expertise in understanding and managing contracts in detail is essential.
- The rapid *development of digital technologies* into essential tools for increasing both corporate and project efficiency. This has created the need for specialised roles for CTO's and IT specialists to implement and manage these programmes, such as building information modelling (BIM), intelligent document format (IDF), robotics, 4D reality models, etc. (refer to Chapters 20–22).
- The need for a really good understanding and acceptance of *the importance of human dynamics* by managers at all levels, from department heads and CEO's all the way down the chain of responsibility to site managers. HR managers have an expanding role in this, in establishing policies and advising management and employees in respect of personal behaviours, effective communication, collaboration, and relationship management.

It is now generally accepted by progressive managers in many industries that 'emotional intelligence' (acronym EQ) is an important element when evaluating the soft skills of personnel or potential employees. Soft skills are defined as personal attributes that enable someone to interact effectively and harmoniously with other people.

Global Construction Success, First Edition. Charles O'Neil.
© 2019 John Wiley & Sons Ltd. Published 2019 by John Wiley & Sons Ltd.

It is also generally recognised that it is more difficult for people to master soft skills/emotional intelligence than technical skills, but that soft skills can be game changers in terms of your people and your projects. Given the impact of human dynamics in so many areas of construction it is particularly important that our industry leaders, senior managers and project managers possess these soft skills.

In Chapter 10, Tony Llewellyn explores a very important aspect of this in respect of construction and shows why social intelligence is a critical ingredient to project success

In summary, construction contractors and government procurement departments that are not being proactive and progressive in addressing these four areas will continue to be exposed to having projects that run over budget and programme, and be at risk of incurring serious project and corporate losses.

11.2   The Argument for Broader Based Training of Tomorrow's Industry Leaders

As an example, take the situation of bright young project managers who start on site and whose ambition is to rise through the ranks to senior corporate management. I believe they should have a broad understanding and some working knowledge of all the key elements involved in running a project, but obviously they cannot have a specialised knowledge of all the main areas. These key elements include planning and approvals, feasibilities and financing, programming, estimating, cost planning and risk management, the structure of contracts, processes and systems, and last but not least the construction methodologies that they are managing (they are likely to have building or engineering degrees anyway).

The idea is that this gives them a good understanding of the bigger picture and enables them to communicate with the specialists from each discipline. My training in Australia included a four year technical course in building and construction, then a specialist course in planning and programming, followed by further training in estimating and cost control. We had to be a jack-of-all-trades if we wanted to progress. If a colleague couldn't do something for some reason, such as they were on holiday, off sick, or snowed under, then you pitched in and did it for them.

In my view, construction companies that encourage and nurture this approach will end up with quality leaders who will provide sound foundations for project success and company growth.

11.3   What Makes a Good Leader in the Construction Industry – for Contractors, Government Departments and PPP Players?

There has been much written on the credentials of good business leaders, but the following list is what I have observed in this industry over the years:

- Honesty and integrity is a must
- A straight talker, no hidden agendas

- Has an in-depth knowledge of the business and its place in the market
- Provides clear decisive direction, based on sound strategic thinking
- Has a sound knowledge of financial management, with a conservative approach
- Delegates well and does not micro-manage; not stressed; has time for 'thinking'
- Well organised time-wise, leaving time to meet people and visit projects
- A good communicator; a pleasant personality; has presence and commands respect
- Listens, asks for and heeds advice
- Appoints the right people to the right job, no cronyism
- Appreciates performance and easily distinguishes people who are 'all talk'
- Promotes training and further education; up-to-date with technology
- In summary, knows what is really going on in all aspects of the business and manages the bigger picture.

Unfortunately big talkers with big egos often get themselves into senior positions, only to be found out later that they do not have the all-round skills to do the job. This is where Board directors have an important role to play, to ensure that suitable people are appointed.

11.4   Personnel Recruitment and Positioning – A Different Perspective

If you accept my contention that the human aspects of project management are equally or more important than technical skills, then it follows that personal or soft skills should carry significant weight when recruiting personnel, whether for long-term employment or short term for a specific project.

I developed the following interview checklist and used it for many years. You will note that it is heavily weighted towards character, personality and personal skills rather than technical skills. By the time an applicant reaches the shortlist interviews for a job, it is more than likely that their technical skills have been properly checked out and are acceptable, but even if they check out really well it does not necessarily mean this person will be a good team player, be a pleasure to work with or fit into the organisation well.

This checklist should not be tabled during the interview, because that would be too much 'in your face', but make notes on it immediately afterwards as appropriate.

Of course, no one is perfect and allowances nearly always need to be made in respect of some of the personal traits that all of us have.

	Interview Checklist 'Fit the Team' Criteria	Comments
1	Experience – specific construction/engineering/PPP	
2	Experience – broad enough and suitable for our company	
3	Experience – specialised enough for the position	
4	Experience – hands-on or academic	
5	Presence generally	
6	Presence at high level	
7	Uncomplicated personality	

	Interview Checklist 'Fit the Team' Criteria	Comments
8	Eye contact	
9	Open and forthcoming	
10	Listen – and hears – attentive	
11	Talks too much	
12	Talks over the top	
13	Exaggerates	
14	Understates	
15	Alert/sharp witted/street smart	
16	Dress and shoes – neat or sloppy	
17	Knowledge of our company	
18	Energetic/workaholic or 9.00 to 5.00 type	
19	Is this person only looking for a fill-in position	
20	Recreational pursuits	
21	Why moving now/why previous moves	
22	References (for what they are worth)	
23	Work culture expectations	
24	Salary/package expectations	
25	Other interviews recently	
26	Nervousness/adrenalin running (not necessarily a negative)	
27	Overall perception	
28	Would I like to work with this person	
29	Will he/she really fit in and add value to our team	

Interviewer: remember that no one is perfect!
Interviewee: 'You never get a second chance to make a good first impression!'

(Will Rogers)

11.5 Leadership Considerations

11.5.1 Real Team Leaders versus Egos, Arrogance, and Poor Basic Management Skills

It is not uncommon to find arrogant managers who fuel their own egos by putting down their staff. This can be very disturbing and stressful for the staff and leaves individuals wondering about their own capabilities so it is important to not be deterred if you think you are right and have not performed badly. Be patient and sit it out, perhaps talk about it with your peers or human resources manager, as your day will come if you stay focused on doing your job properly – and so will theirs! Their behaviour is often a cover-up for their inadequacies as managers.

Amazingly, it is also not uncommon that some of these arrogant types also treat personnel from joint venture partners, and even clients, with rudeness, superiority and disdain, but generally they don't get away with it for long. A few large corporations over the years have fostered this type of behaviour as a company culture, but they are not doing themselves any favours in respect of developing business relationships. It should not be tolerated.

A complete opposite to the above type of manager is the confident and unassuming manager who appoints the best possible people to their team on the basis that their high performance will make the manager's life and workload easier and help them get optimum results, thereby also making them look better in the eyes of those above. There is always the risk that in doing this someone you appoint will outshine you and end up taking your job, but it is a small risk and still worth doing for the sake of the company. Everyone reaches their level of incompetence sooner or later, but it is hard to recognise it in yourself and probably even harder to accept, but if it is your turn then that's life!

11.5.2 Cronyism

How often do you see it and how rarely does it work – *'he is a good guy and I can trust him; he is really bright so he will learn how to handle it quickly'*, which actually means *'he is a close friend in need of a job and he will do what I tell him without asking awkward questions or rocking the boat'*. This is traditional construction machismo and is totally out-of-date.

11.5.3   Bosses with Poor People Skills Who Avoid Staff Management Problems

Another significant factor that can be detrimental to good collaboration, communications and staff performance is that too many senior managers have poor people management skills and are just focused on operational matters and business development activities. This is common in all organisations and government departments, not just with construction companies.

So, typically, you get some alpha types who might be the 'red' assertive, dominating personality, or the *'blue'* detailed perfectionist (see Chapter 7), and they might be pretty hard to handle, have personality clashes, etc.

If a boss with poor people management skills finds this area too hard, he or she will just bury the problem or ask the HR manager to sort it out, which rarely works. Or worse, the boss might play off one person against the other, thinking this will get the most out of both.

This is a big mistake and often results in the loss of very capable people, who get frustrated that their staff issues are not being addressed by the only person with the authority to sort them out – the boss.

A boss who is really 'aware' and has good insight as to what is happening in this regard will sort it out positively or seek some advice. Really switched-on bosses with their egos under control might also undertake external coaching in this area if they feel it is necessary.

After four months of negotiations, the terms of sale of three Australian hotels to a Japanese investor were finalised. Three Japanese executives arrived in Sydney on an overnight flight to formally sign the sales contract at a scheduled 9.00 a.m. meeting. After polite greetings and introductions the contract documents were tabled for signature, with the principal Australian signatory being the MD of the hotels division, owned by a larger conglomerate. At this moment the MD's mobile phone rang and he took the call (it should have been switched off). He stood up and walked over to the window overlooking the harbour and talked and talked and talked. Three times his colleagues tried to interrupt him, but he brushed them off. Finally, after 15 minutes he returned to the Board table and just said *'let's get on with it, I have a busy day'*. The Japanese gentlemen, who were clearly annoyed, had a quick discussion, said they wished to defer the signing of the contract and departed. They sent a polite note later that day withdrawing from the sale.

The MD's behaviour would have been extraordinarily rude and disrespectful to any nationality, but this MD had actually lived in Japan in a senior executive position for a number of years. He lost his job soon thereafter.

Comment: we live in an international world with multi-ethnic populations in most countries and it is completely unacceptable to have managers or any personnel who are not respectful of other nationalities and cultures. If someone has a problem in this respect then they should be asked to step aside and let someone else handle it, because their behaviour is likely to be hugely damaging.

11.6   The Inherent Risks of Decision Making for Survival

It is entirely human and understandable that people at all levels of an organisation will make different decisions if their job security is threatened.

This may be done consciously or subconsciously, and to different degrees of significance in terms of the overall risks being taken with a project, but one thing is for sure, when decisions are made on this basis then additional project risk is an inherent factor.

Let us look at a simple example. A construction company (the contractor) is under pressure from its bankers and the word is around that *'if we don't win these next two or three projects then it is possible the company will go under'*.

So the Bid process for these projects commences and there is immediate pressure from above to subcontractors and suppliers to cut their prices to the bone. Several of these may be owed substantial amounts of money by the contractor and payments have gone out beyond 120 days, which means that the contractor is relying on new projects to sustain cash flow, sometimes referred to by bankers as *'overtrading'*.

The wise thing for the suppliers and subcontractors to do would be to crystallise their situation right then, possibly accepting losses, and not participate, but this depends on how much real information they have on the contractor's financial state; on how large and significant to their business is the money that they are owed; and on the relationships and loyalties they have with the contractor.

So they go ahead and give very competitive quotes. These reach the estimators, planners, and mid-level managers. They are all worried about their job security and home mortgages. Even with the best professional and objective assessment it is likely that consciously or subconsciously they will now make decisions that they otherwise would not have made if the contractor was in a sound financial position.

The Bid proposals reach senior management and the executive directors. Most likely they have a Bid review committee. Their situation is probably no different to those lower down the pecking order; maybe it is even worse because of their shareholdings, higher salaries, and their age insofar as getting another job at a similar level of remuneration will be difficult, especially if they are tainted with the collapse of their previous company.

So the Bid review committee takes the view that they will reduce the contingency sums, risk allowances and margins, and rely on doing the project efficiently, making some extra margin out of the variations, and from further 'screwing' the suppliers and subcontractors during the course of the project. In reality they are compounding the sharp pencil effect that has already taken place lower down the line.

The finance director asks how on earth they can have a positive cash flow on this basis when they still have to provide performance guarantees and accept the retention sum deductions along the way. They are told that things are going to improve for the contractor generally if they keep winning Bids and the directors will worry about the cash flow situation down the track, after all, look at the good *'front-end loading'* they are building into the milestone progress payment schedule!

So the contractor wins two new projects on this basis and they commence work. Often, this is when Murphy's law comes into play and the rest is history (see also Chapter 4, items 8 and 9 of the list of causes of failure).

11.7   The Human Fallout from a Failed Project

When projects turn sour the impact can be quite devastating to a wide range of stakeholders and individuals. On the client side the people that conceived and ran the project will likely pay the price, maybe with serious consequences to their careers, unless it is the boss of a private company that made all the wrong decisions, in which case it is their personal pocket that is hurt.

However, that does not prevent contractors, subcontractors, suppliers and individuals further down the chain from suffering the financial consequences. The most evident damage is mostly seen when a head contractor is terminated for poor performance, has major losses, or even goes to the wall on a series of bad projects, which is not uncommon with small/medium size contractors.

In this situation the people that get hurt most will be the professional consultants, subcontractors, and suppliers. Lenders, investors, large construction companies and government authorities give no mercy to the smaller players when these big guys get into trouble, with the result that some of these smaller players may go broke. These larger entities try and protect themselves with specific contractual agreements tailored for such eventualities, such as nonrecourse agreements and indemnities, but they also commonly write off large sums related to failed projects.

When the smaller businesses suffer losses or go bust as a result of their involvement in a major project that has failed, this often results in job losses for many genuine, hard-working people who are far removed from the corporate decision makers who really caused the problems. These business and job losses can in turn result in people losing their homes and other personal assets, and being declared bankrupt.

It is also a well-known fact that unscrupulous property developers and construction head contractors all around the world enter into multiple subcontracts with professional consultants, subcontract specialist trades and suppliers with the specific intent of finding reasons to never pay the last 10% of the subcontract value. Small under-capitalised businesses operating on fine margins cannot survive this for very long, but the unscrupulous bad guys know that there are invariably new subcontractors willing to work for them and take the risk. It is a tough industry in this regard.

Fortunately, this unsavoury practice has drawn the attention of legislators in an increasing number of countries, such as in Australia where the various states have in recent years introduced *'Security of Payments'* legislation, which is quite effective in protecting subcontractors from unscrupulous employers.

11.8   Summary

'People and teamwork' are one of the four main themes of this book. Having the right team led competently by the right managers is critical to both project and corporate success.

The objective of this chapter has been to point out the absolute necessity for effective leaders to have the full range of business, technical and people management skills, and for managers at all levels to be fully competent to handle their respective responsibilities and to participate as conscientious members of the bigger team in delivering this success.

Section C – The Right Framework – Forms of Contract, Business Models, and Public Private Partnerships

12

The Contract as the Primary Risk Management Tool

Rob Horne

Head of Construction Risk at Osborne Clarke LLP, London, UK

When does the risk start to accumulate in a construction project? Is it once an unexpected problem has been encountered? Perhaps earlier, when processes are put in place for allocating responsibility for what happens on a project? Maybe before process, when an agreement is reached on what the project is and who will do what to complete it? Perhaps even earlier, when the project is first conceived? The reality is construction operations are inherently risky. To manage the inherent risk there needs to be a system or process in place. In a very general sense that is what the contract is there to achieve. The contract is therefore the first and primary risk management tool.

Most disputes in construction projects revolve around how well the contractual documentation, drafted at the outset before work has even begun on site, has reacted to the challenges and inherent risks of the project. Therefore, this chapter is not an explanation of what is 'bad drafting' in the sense of sentence structure and choice of words or what particular clauses should or shouldn't be included in a contract to make it effective[1]. Rather, this chapter is going to consider the principles that should guide the drafting and do that from looking backwards from a dispute or issues occurring on a project to its root cause in a risk management sense.

Far too often in situations where dispute or conflict feels inevitable the apt, though overused, phrase *'I wouldn't start from here'* can be heard. While this may be apt, and indeed is almost certainly a truism, it helps very little with the problem the parties to a project find themselves in. However, following the thread back through the life of the project can lead to some surprising insights into the root of the problem. What may have started as a simple, well intentioned, innocent looking piece of drafting in the Boardroom[2] can looking remarkably different midway through a project. Looked at from the project end of the telescope effective use of the contract as a risk management tool is often viewed very differently to how it was viewed while being prepared and agreed.

Equally, it is tempting, when considering contract drafting, to consider only the lead construction contract. However, as will become obvious on analysis, this is far too narrow a focus. The whole supply network needs to be considered from employer

---

1 Although I do conduct training on the meaning and application of certain standard forms and I am often asked for my view on the effectiveness of certain terms and provisions within the overall drafting exercise.
2 Throughout I use 'Boardroom' for the senior management of the contracting parties whose job it is to enter into the project documentation but not carry out the detailed delivery of the project.

Global Construction Success, First Edition. Charles O'Neil.
© 2019 John Wiley & Sons Ltd. Published 2019 by John Wiley & Sons Ltd.

to designer, to subcontractor to supplier. Any break in the commonality of the risk management approach within the supply network will increase, rather than minimise, the chance of risks crystallising to form disputes[3].

While it might be tempting to look just to the drafting of the words on the page, point the finger, and says *'there's your problem; you wrote X in your agreement when you meant Y'* that is, regrettably, a vast over simplification. Who wrote those words? Why were they written that way? How do they fit in with the rest of the project documentation?[4] How are they understood at a project as well as at Boardroom level? Stopping at the words on the page, while certainly what a lawyer will do to reach a legal answer to a specific question, is not the route to solving the *'don't start from here'* dilemma and neither is it the mechanism through which problems on future projects can be solved. Further, it will not of course even begin to deal with perhaps the greater evil, and almost always the *'elephant in the room'*, that nobody thought to look to the contract until there was a problem, by which time it was probably too late to minimise and avoid the problem, and the parties were forced to resolve it.

While this chapter may be an exploration of the idea of using the contract to actively manage project risks it is within a much larger context of considering and understanding the human dynamic that underpins and drives projects from inception, through drafting to delivery.

The written words are meaningless without human interaction to bring them to life and give them meaning. In particular, the words have to be operated and adhered to, otherwise, no matter how good (or indeed bad) the drafting, it will play no part in minimising the risk of a dispute. Therefore, while going through some common issues that arise, the human input driving the potential problem, dispute and solution should become apparent.

A good place to start is with a hypothesis; you may agree or not – the purpose is to shape the discussion and deepen the understanding. So the hypothesis here is that there are just six key or principle issues that drive problems using the contract as a risk management tool. Those six are like the ingredients of a cake, if one is missing or not quite right the cake as a whole won't be quite right, although it will still bear many of the qualities of the cake you were expecting. It can be difficult to identify what was missing unless you have significant expertise; however the result is plain to see.

The six principles, or ingredients, to creating a contract that can be used as an effective risk management tool are:

12.1 Common understanding (or lack thereof): in particular with regard to project outputs – what a successful project will look like, and project inputs – what is needed to create the output. The lack of understanding could be on the part of any or all of the project parties and the further out of alignment the understanding is the more likely a dispute is to occur. Recording this common understanding is one of the key purposes of a contract but one that often gets overtaken by the detail of specifications and the like.

12.2 Clarity: having understood what the project is about, the way that understanding is recorded needs to be clear and unambiguous. This is the issue that is most

---

3 There are very few true multi-party construction contracts, but see the PPC2000 form of contract as an example of such an approach.
4 I do not say the contract documents here as that would be over simplifying the background to how contract documents are arrived at and is a topic I shall return to.

readily associated with drafting, often to the exclusion of the other principles identified and described here, leading to over simplification and a significantly reduced chance to avoid the 'don't start from here' scenario. Clarity does come from appropriate use of words and sentence structures but if it is not properly linked back to the first principle, common understanding, then it will inevitably fail to be clear no matter how good the individual words used.

12.3   Knowledge transfer: it would be nice to think that a set of words can mean the same thing to all those who read them[5] Unfortunately, experience shows that this is not the case. Even where there has been good understanding and clarity at Boardroom level, when it comes to detailed implementation of the project that knowledge needs to be transferred effectively. Therefore, a key part of the drafting exercise, if it is to minimise disputes, is its transmission from Boardroom theory to site operation.

12.4   Adaptability: every project is different and bespoke in some way, some more so than others, and major projects are often completely unique, one of a kind, builds stretching the boundaries of engineering and architecture[6] There are a great many unforeseen and even unforeseeable events that occur on construction projects and therefore, layered on top of the understanding, clarity and knowledge transfer must be an element of adaptability. This is one of the great balancing acts; achieving adaptability without taking away from clarity.

12.5   Acceptance: one way or another, whatever the contract says needs to be accepted. It is extremely unhelpful, but all too common, for one party, during the construction phase, to think to itself (and on accession, say to the other party) *'these terms are not fair, we need to change them or abandon them'*. Such wholesale changes can be extremely effective but only if there is an acceptance of the original terms and therefore who is making what concession.

12.6   Application: the final principle here is that the terms must actually be applied and adhered to. There are innumerable projects, and indeed construction professionals when considering the human dynamic, that rely on putting the contract 'in the drawer' and only retrieving it to deal with problems after they have arisen. For the purpose of this discussion (and almost certainly in a general sense as well) this is not application. Proper and timely application must be considered part of drafting as the way in which a contract is written will, to a large extent, define whether it is a useable tool to help guide the parties or a weaponised rulebook only used after the event to try and make a point.

All of these principles, to a greater or lesser extent, can be guided by effective drafting of the contract. The contract is after all, the record of the agreed rights and obligations between the parties and therefore the 'answer' to most, if not all, questions should either be in the contract or capable of inference from the contract. Put simply, for the contract to be an effective risk management tool you have to get it right from the outset. Build the foundations correctly if you want what is built on top of them to last.

How do these key issues manifest themselves in projects? Unfortunately it is all too easy to identify examples of where one or more of the ingredients have been missed.

---

5 This certainly seems to be the utopian world that many lawyers live in.
6 See Chapter 4 giving a history of the Sydney Opera House.

Project Example 1 Middle East Residential Development

Five contractors were employed to construct 200 or so villas each, with a mix of five different villa types. Each contract was carefully negotiated around an International Federation of Consulting Engineers (FIDIC) standard form of contract. A separate contractor was employed to undertake the infrastructure work across the project as a whole and this approach was understood by all of the contractors. There was a fairly typical term included in each of the five villa contractor contracts requiring all of the contractors to work together and coordinate their works.

Halfway through the build some general problems across the development occurred, including sandstorms and a general material shortage, particularly in relation to concrete products. In addition, the infrastructure contractor commenced his work and, unfortunately, because he was the only infrastructure contractor, there was no requirement placed on him to coordinate with others.

The infrastructure contractor therefore identified the best sequence of work for himself and commenced. This had obvious knock-on consequences for the villa contractors as they could not access certain areas of the site due to roadways being excavated to install drainage and other services. The shortage of cement based products, because each contractor was under separate contracts and potentially subject to damages for late completion, placed each contractor in competition to secure an adequate share of the small supply. Some contractors managed to secure full delivery while others received almost nothing. The sandstorms in turn made it very difficult to carry out anything other than precast concrete works but the short supply magnified the effects of the storms.

On this project the contracts did not provide the answers the parties really needed and disputes (or at least differences) arose. Looking back at the principles identified in this chapter, the contracts for this project had reasonable common understanding of the nature of the project but a key element was missing, that a very important factor was going to be the sale of the villas as they became available to provide revenue to the developer.

The contracts were not sufficiently clear, particularly in relation to the interface between the contractors. Knowledge transfer occurred but, because of the limited common understanding and clarity, this was inadequate to meet the demands and challenges of the project. The contracts, as drafted, were insufficiently adaptable to deal with the changing circumstances of the project that, while extreme, could have been foreseen as possibilities in a project of this complexity. There was little acceptance of the terms of the contracts once the circumstances described occurred. What ensued was a general 'blame game' where all parties chose to read their contract in a different way, no matter how strained the interpretation, in order to try to achieve their commercial goal.

Finally, there was insufficient application until the problems were manifest and the parties were reliant on the contract to justify and support the actions they had already taken rather than it being used to formulate and regulate the relationship before either party acted.

When, then, can parties take steps to improve the contract for use as a risk management tool? The most obvious starting point (although still lost on some projects) is before they start. If you have already commenced and then try to structure the contract around what you are already doing the risk exposure of the parties is markedly higher. There are, however, occasions when the drafting can be improved after the project has been started, but I will return to this shortly as it is easier, and more beneficial, to consider the opportunities in sequence.

Step 1 – Project Inception. This is the very earliest stage at which the kernel of an idea for a project is formed. You could think of it as a *'light-bulb'* moment where someone says *'hey, I've got this great idea… '*. However, I prefer to think of it as a 'big bang' moment. The key difference being that in a *'big bang'* moment, like the formation of the universe, everything about the project in fact crystallises at that moment. You may not realise it, understand it or deal with it yet, but all risks and opportunities coalesce in that *'big bang'* moment. Soon after the initial big bang, ideas will start to be reduced to writing and this is the protean form of contract. It is at this stage that the first cut of opportunities and risks will be considered and it is at this stage that the first of the principles described above is at the fore. If unrealistic expectations are to be avoided, right at this very early stage, risks and opportunities need to be considered and how the intended contract is going to deal with them understood.

The common understanding will probably start from a set of outputs from the project: what is it we are trying to achieve from this project? Why is this particular formulation better than another? Even though the detail of how those objectives are to be carried into effect will change and develop, properly recording the original objectives is key. With the proper initial objectives recorded, development of implementation can be measured against something and there will be an *'essential quality'* of the project around which a common understanding can grow and allow the other principles to be achieved.

Project Example 2 Cambridgeshire Guided Busway

This is a unique piece of infrastructure, being a specialist and dedicated running track for buses only. In Cambridgeshire the route of the busway was to follow an existing, but by that time, disused railway line from the town of St Ives to the edge of the City of Cambridge where the buses would transition onto bus lanes on conventional roads, and then back onto a dedicated guideway to link to the railway station for Cambridge and Addenbrookes Hospital.

The *'big bang'* moment was the need for alternative transport into the City of Cambridge and was part of a multimodal study for all transport in and around Cambridge. The big bang reduced into key objectives around the speed of transportation, ride quality, and accessibility. Consideration was given to numerous ways of achieving these objectives, including renewing the railway, constructing a conventional but dedicated road, using a slip-formed concrete guideway or using a precast solution. As the objectives of the route had been thought through at project inception the options were narrowed to a precast concrete guideway and there was an easy and obvious common understanding on the key priorities for the route.

As history has shown, from resultant disputes, particularly with regard to this second project example above, clarity, common understanding and all the other principles being in place at project inception is no guarantee that the project will be dispute free. Equally, the work done at project inception will not amount to effective drafting to manage risk and minimise disputes; it is just one step one the way to achieve that. There are, at project inception, still many stages at which the contract can become ineffective as a risk management tool or in fact can positively promote and encourage dispute.

The knowledge transfer and other drafting issues at this stage are focused on drawing in the professional team to the developer. Beyond project inception, the developer usually takes a step back from the project while the professional team start to add detail. Therefore, if the essential objectives of the project have not been properly recorded, or drafted, at project inception stage the detail of the project will be layered on top of an inadequate foundation and is therefore far more likely to lead to a misunderstanding of the inherent risks, preventing proper management of them and leading to disputes.

Step 2 – Procurement and Tender. Where the project inception step was all about initial ideas and recording them to ensure focus this step is focused on the initial contact with the team that will take the idea to reality. The key part of this step is that it will define the expectations of both parties. The principles at the forefront here are the creation of a common understanding, now between developer, professional team and constructor, and clarity of those documents that are going to be defining those expectations and requirements.

The temptation of many developers is to hurry through this stage, almost as if it is a necessary evil rather than an important part of the process. The rise, to dominance in many areas, of design and build construction has led to many developers seeing the tender stage as an expenditure to be minimised or avoided as the risk and requirements of bringing the project into reality are going to be passed to the constructor. The most often encountered problem with this is that, once in the build phase, the developer does not like some detail of what is being constructed and therefore seeks to change it. Many developers omit to consider or do not accept that such changes may well lead to unexpected consequences such as additional cost or time to complete.

Project Example 3   Northern Ireland Housing Executive v Healthy Buildings Limited

This project experienced a dispute that ended in the Northern Ireland Court of Appeal.

Healthy Buildings were appointed by the Housing Executive to carry out asbestos surveys on various numbers of its housing stock that were being regenerated. The appointment contract required Healthy Building to comply with relevant legislation in the carrying out of the investigations and made particular reference to the standard requirements for how investigations were to be carried out. However, the standard allowed for a choice of two different ways of carrying out those surveys; one amounted to a survey by example and the other amounted to a detailed survey of each house to be regenerated. The Housing Executive did not specify which method they preferred or required at tender stage. Just after the contract had been awarded, the parties met to discuss implementation. At that stage the Housing Executive made a selection of the method of survey they wanted, preferring the more detailed house by house option.

Unfortunately, as the need had not been communicated clearly and there was no common understanding, Healthy Buildings had priced on the cheaper option. The Housing Executive did not want to pay more as they said it was open to them to choose either option as both had been in the tender documents. Healthy Buildings did not want to carry what would be a considerable loss as they had been led to believe the choice was theirs.

The court found that the choice of one option over another was a change and therefore the Housing Executive was to pay the extra cost. The lack of clarity had left the choice in the hands of Healthy Buildings rather than the Housing Executive.

This step in formulating the eventual contract should add to, strengthen and support the foundation stone laid during the project inception stage. Imagine two foundation blocks working in tandem to support a building; as long as they work in tandem everything that follows should be simple. However, if they are misaligned so they do not produce a single level starting point, or perhaps if they are made of different materials that will react differently when stress is applied to them, then the opportunity for a problem has been created. The further out of alignment they are the harder it is to recover, the more substantial the difference in material (even if both materials can be used as a proper foundation on their own) the more likely it is that additional and unexpected stresses will be applied during the project.

Step 3 – Contract Negotiations. The discussions and negotiations that occur around and during the tender period allow alternative views to be put and an opportunity is created for additional benefit to be added by the constructor. The increased use of two stage tendering, where one stage is competitive and then there is a second stage with a preferred bidder where more ideas and opportunities can be explored, certainly promotes the importance of this stage[7].

The key principle at play during this stage is clarity with a healthy dose of knowledge transfer and a deepening of the common understanding. To achieve these three principles the discussions and drafting being produced at this stage needs to be done in as open a way as possible if disputes are to be avoided. Unfortunately, it is often at exactly this stage that the paths of constructor, developer and professional team can easily diverge.

During this step everything is starting to get quite 'real' and the tendency for many parties is to look to their own interests first. Indeed, why wouldn't that be the case? The developer will either have public accountability if it is public body or will be answerable to shareholders if private. Equally, the constructor, who will probably be working on a very tight margin,[8] is carrying out the works for reward, in other words he needs to make a profit. The professional team will have significant influence and control over the constructor but will not, generally, be answerable to the constructor. The natural instinct of all parties is to protect their own interests first. In doing so, that protection will often not be through an open dialogue until each party think they understand the risk they are taking and look for an opportunity to minimise the risk they carry and maximise the opportunity from risk carried by others.

It is rare for there to be a fully open joint examination of the risk inherent in the project as a whole, such discussions, when they occur, tend to focus on specific aspect of the project or particular issues which have arisen during the discussions to that point. If one were to look for a single point at which the preparation and drafting of contracts most often goes astray or stops being about minimising the chance of future problems, it is this step in the process.

---

7 For additional information see, for example, *Early Contractor Involvement in Building Procurement: Contracts, Partnering and Project Management* by David Mosey.
8 A profit margin, and therefore risk margin, for a general contracting company in the UK could be expected to be from 1% to 5% and that profit margin will usually be expected to absorb any risk the constructor cannot mitigate while on site.

Project Example 4 Design Appointment on a Design and Build Project

This was a project for the construction of a road in the UK. A fairly challenging timetable was set but the constructor had appointed a designer to assist it during the tender stage. The contractor won the work on the basis of a lump sum price and fixed completion date set out in a detailed programme.

The detailed construction programme included dates for design release and activities for carrying out work on site following the release of those designs. The activities flowed logically through the project from one to another through to completion within the contractual limit. The logic within the programme linking activities together and creating a dynamic model for the project was sound and allowed for good management of the activities and clear forecasting of progress.

What was missing was support to that programme from others in the supply network, in particular the designer. The designer had provided confirmation that it would provide design in order to allow the activities in the constructor's programme to be achieved. While this was acceptable as a statement it did not provide sufficient clarity to anyone on the project once site activities commenced and sequencing started to change. What became apparent very early on in the project was that the constructor was not sufficiently managing the designer to ensure designs were available, in an approved state, in time to allow construction to proceed without interruption.

There was clearly a good common understanding of what was needed but clarity and knowledge transfer had been lost. This cascaded into a failure by the parties to accept the very simple design programme intent and removed any adaptability from the build and from the contract. Despite having a *'partnering charter'* signed by all of the parties at the outset of the project it was soon clear that the failures in drafting had led to a problem and a dispute followed over whether the designer had caused delay to the project.

Step 4 – Contract Execution. Although many would see this as not really a step in the drafting of a contract it is a crucial moment in terms of enabling the risk management aspects of the contract for a number of reasons. First, it is the last chance to get things right in the drafting before rights and responsibilities become locked. Second, this is the key point at which knowledge transfer comes to the fore as the lawyers (and often the entire Bid team for both parties) will step away from the contract at this point and leave it to the parties (and potentially an entirely different delivery team) to implement it. Finally, there is a real danger that this is the step at which the parties believe that the contract has been *'locked'* and therefore they can put it away and not look at it again unless there is a problem, at which point they imagine it will, miraculously, provide an answer acceptable to everyone. Rather than a risk management tool it is reformulated to act as a disaster recovery manual. Prevention is always better than cure.

Most of the Private Finance Initiative (PFI) or Public Private Partnership (PPP) projects I have had any involvement with tend to run contract execution up to 11.50 p.m. or so on the day by which contracts must be signed or the funding arrangements will fall away or some equally dire consequences occur. In those last hours there is some very hard, close scrutiny of the project documentation. However, it is highly unlikely that the focus of attention in such projects is on the risky part of actually building something; far more often the key focus is the functioning of the financial model. That is not necessarily a bad thing, I raise the point because the financial model is the real driving force behind the whole project and yet it is being considered, in detail, right up to the last possible moment. When you compare and contrast that to the position in a normal, employer funded project it is unlikely that such focus is brought to bear on the relevant construction documents. Either they are agreed well in advance or, just as often, they are left, with work commencing under a letter of intent on the assumption that the construction contract will be entered into at some point. With such lack of attention one cannot blame the drafting if there is a problem later; it was never really given a chance.

Perhaps of greater significance however to the project outcome is what happens in the immediate aftermath of contract execution. Most of those involved with drafting the contract, preparing and submitting tender documents and negotiating on particular issues will step away. It is at this moment that many problems with the contract drafting will crystallise. Those problems are not necessarily manifest problems with the drafting itself, but simply that the way it has been considered and approached during the tender and up to execution is re-reviewed as the implementation of the project starts and the theory of what, until then, has been a sterile contract is put into practice. The knowledge transfer that needs to occur at this point is critical to success. You may think this has little to do with the contract drafting and, in a traditional sense, that might be correct. However, challenging the traditional approach, why isn't the contract drafted properly in the expectation that it is to be implemented and applied by others[9]?

Step 5 – Build Phase Implementation (stress testing). Once a contract has been entered into and fully executed, its function and importance as a risk management tool can be assessed under 'live fire' conditions. There are a further two key aspects to consider, and at this step the principles of adaptability and acceptance become key.

The first of the two key aspects is the stress testing of the words written into the contract, how they are used and followed, and how well they stand up to the scrutiny and challenge of a live project with changing facts and circumstances. The stress testing will focus significantly on the principles of adaptability and acceptance. However, this step can be crippled before it begins if common understanding, clarity, and knowledge transfer have not been put in place through earlier stages of the drafting process. In other words, for the contract to act as a proper risk management tool, all of the ground work must have been done, from project inception up to start of works on site.

---

9 This is the precise point made by the NEC for the form of contract that is deliberately drafted for implementation on site by engineers rather than for legal review by lawyers. This creates a second tension as the contract still needs to be legally enforceable, but it is perhaps a step in the right direction to properly appreciate the reality of how the contract is used.

Project Example 5 A PFI Project in the UK

This was a project for the construction of a new hospital with an SPV contracted to the local health authority and a build contractor carrying out the initial work (owned by the main equity holder in the SPV) and a facility management (FM) contractor (that had no interest in the SPV).

During the course of the build phase there were various changes to the room layouts and makeup as well as to how the services were to be provided. The changes being implemented were relatively modest in the build phase but, through the choice of equipment being made by the build contractor, had very substantial repercussions in the operation phase. The majority of the additional build phase costs were fully recoverable whereas the majority of the operational phase costs would end up being borne by the SPV.

The various contracts, despite being of significant size and complexity, did not provide for this scenario, and neither was there sufficient flexibility with the various contract documents to react adequately to this new challenge.

In fact, in this project the problem was solved by the SPV equity holder acquiring the FM contractor and therefore being able to balance build and operational risk across the whole project. This is a good example of the human dynamic working to solve a problem.

The second key aspect is change. Practically every construction project will have a degree of change in it; it is an inherent part of the construction process. Whenever change occurs, how that change is recorded and managed should be approached with the same care and diligence as the original contract drafting. Certainly the same principles as outlined above in relation to the contract itself will be relevant.

Unfortunately, it is rare for the same care and attention to be given to change as to the original contract concept, even where the change itself is very significant. The key issue here is that a change in the nature of content of the works can alter the underlying risk and therefore the risk management tools of the contract may become less effective. In reality, while it may look and feel easier and quicker to implement change during the build phase of a project the risk management of the project as a whole must be considered when doing so. It may well be that new tools and new controls need to be brought in to help support the project where change has been needed. Is the change, for example, a symptom of some part of the contract not working properly? If it is, how do you trace back from the symptom to the cause? The only way is diagnosis, through good commercial intervention, and using the tool that the contract has given you to examine what didn't work properly. An answer, which is heard all too often, of *'we just decided to…add works…change the sequence…alter the design…, etc.'* is not acceptable if one actually wants to learn to manage risk. If one is in the business of having disputes or perhaps playing roulette with millions at stake then perhaps it is fine.

Project Example 6   A Road Project in the Middle East

A project was let for the dualisation of a length of single carriageway road. The project was tendered and a contractor selected and appointed under an amended FIDIC Red Book form of contract. A programme was agreed and a lump sum price was put in place. Within a few months of contract award (on a 30 month project) a variation order one was issued. Variation order one provided 'add one further lane in each direction to provide a three lane motorway for the length of the project'.

With such a significant change the contractor provided a detailed cost and time quotation for this variation. That quote was rejected, but with no alternative suggested. Under the contract the contractor was required to comply with the variation order instructing additional work but did not know the recovery he would make for the additional work. The contract, in any meaningful sense, ceased to operate at that point as the time and money provisions had, effectively, been set aside. The parties continued to negotiate the time and cost implications of the first variation order, issued soon after contract award, all the way through to project completion.

These are the five steps or opportunities in which to create a risk management tool, in the form of a contract, to minimise the chances of dispute and the six principles to consider at each stage, as well as during the transition from one stage to the next. However, is that really enough to understand how to get an effective risk management tool in place? While it is tempting to say *'no there are always more chances'* the reality is that it is and the contract must be resolved at the outset. There is little or no point trying to consider what exact form of words will work and what forms will not if the underlying principles have not been adequately considered and brought through the contract to be used by the parties. While examples can be given of specific instances in which certain terms were inadequate it is likely that there have been many occasions where the same, or very similar, terms have been used without a problem. Therefore, it is not so much the words and terms themselves but the underlying principles and their application that is key.

Returning to the main theme, it is not the particular words that are written down that give rise to the greatest problems. It is how those words are identified and then implemented. This then becomes the interface point between creating a principled risk management tool within the contract and the human dynamics of actually applying and utilising those tools.

There is no single right way to draft a contract to make it an effective risk management tool. There is no golden rule that if applied and obeyed will render a project devoid of disputes or allow risks to be identified early and answered in every conceivable circumstance. This is not surprising as, turning back to the start of this chapter, the proposition was that it is not the words themselves that create the problem and therefore it would be strange indeed if, not having caused a problem, those same words can act as a miraculous panacea.

The essence of using the contract as an effective risk management tool to avoid disputes then is in understanding the human dynamics driving the drafting. How is that interaction and knowledge transferred from the Boardroom and the legal and tender/procurement teams to the project team is key to identifying the risks and building relevant management tools around them. How those tools are then implemented at site to turn words on a page into a functioning project is the real test. The contract, whatever words are chosen, needs to reflect the requirements of the team implementing the project. The contract needs to be very carefully considered, usually taking the more time the better, but it needs to be in the context of six principles across five stages, which is as good a starting point as any. Utilising these tools and techniques gives you the best chance of putting in place risk management tools that will provide far more value than the cost expended in considering them and incorporating them into the contract.

13

The New Engineering Contract (NEC) Interface with Early Warning Systems and Collaboration

Richard Bayfield FICE FCIArb

Construction Consultant, Project Director, and Adjudicator, UK

The two themes that are inextricably linked to the New Engineering Contract (NEC) are early warning and collaboration. Indeed, the NEC4 at its launch in June 2017 lauded itself as the 'one, true, collaborative contract'. History will be the real test of NEC4, and whether it does what it says it does on the tin. Put another way, will history show that project delivery under NEC4 gives better value than alternative contract strategies? This is a question that will be better asked in five years' time, once the first tranche of NEC4 contracts have been delivered. NEC4 is now building momentum, with significant early adoption amongst infrastructure clients including £9 billion for Highways England.[1] However for reasons that follow below, the predicted success of NEC4 may not be the best question to ask. I would suggest we first ask 'why early warning?' and 'why collaboration?'

My first experience of an 'early warning' culture and using 'early warning' clauses was whilst building Honda's car manufacturing facility at Swindon in the early 1990s. At this time, the Honda construction team was mandated to introduce lean manufacturing management to its process of construction delivery, because UK construction was perceived by Honda as being 'inefficient'. Moreover, the manufacturing side of the business asked why the construction team could tolerate such inefficiency. Honda quickly moved towards construction management as its delivery mechanism. It understood that it would achieve best value by keeping as much risk as possible within its own portfolio. Early warning was very important to Honda in terms of risk management and therefore important in the strategy for realising commercial opportunity. This meant it was a tool for both threat and opportunity and the clause below was added to the intermediate form as a standard amendment by Honda. The corresponding NEC clause follows below the Honda clause:

> The Contractor shall and the Architect/Contract Administrator may give the other an early warning notice of any matter which in their opinion could increase the Contract Sum, delay completion of the Works, or impair the performance of the Works in use, as soon as either becomes aware of it.
>
> Honda (1991)

1 *NEC Newsletter* (7 March 2018) – www.neccontract.com.

Global Construction Success, First Edition. Charles O'Neil.
© 2019 John Wiley & Sons Ltd. Published 2019 by John Wiley & Sons Ltd.

> The Contractor and the Project Manager give an early warning by notifying the other as soon as either becomes aware of any matter which could increase the total of the prices, delay completion or impair the performance of the Works in use.
>
> NEC1 16.1 (1993)

Early warnings were given within the Honda business that highlighted opportunities to change building design because a new technology had been discovered in Japan (or elsewhere in its global business) that would improve the manufacturing process. Change and flexibility were part of the Honda culture and construction processes. However, around this 'encouragement to change for opportunity' were embedded strong processes to ensure all decisions to change were made from an informed perspective of predicted cost and programme impact.

When the late Sir Michael Latham reported to the industry in 1994,[2] his research recognised the inefficiencies within UK construction. Indeed, he challenged the construction sector to reduce delivery costs by 30%. Sir Michael's research was justified by Honda's second car plant at Swindon where construction costs were reduced by 40%.[3] The savings were achieved through a variety of means including accepting risk, rather than off-loading risk to suppliers, tightly reviewing specifications and not buying a higher standard than was needed, and not including 'future proofing' within the design. The business philosophy was that future changes should be justified by a business case at the time of the proposed change. Finally, collaborative working and close relationships with the supply chain were key to ensuring any errors or misunderstandings were limited to the 'drawing board'.

The measurable beneficial effect of working collaboratively was measured by Honda and included an 'adult to adult' relationship with its supply chain. The benefit of an adult relationship is that information is shared in a collaborative manner, whereas most supply chain contracts are predicated on a 'master–servant' or 'adult–child' relationship.

Back in 2004, the Director of Business Administration at Honda Manufacturing Ltd in the UK. (Mike McEnaney) said something that has long stuck in my memory: '**Claims represent inefficiency** and all benefit by working efficiently'.[4] Subsequently I have had mixed success working in the UK public sector in trying to introduce such thinking. However the prevailing philosophy in many parts of UK construction in 2018 is largely about offloading risk and making sure that 'claims don't occur in my contract'. Whilst this view is explainable (if not understandable) across certain groups, the big infrastructure projects and the big construction spending government departments are missing a huge opportunity by not adopting a 'claims represent inefficiency' approach. If that view prevailed they would work closely with the supply chain to improve efficiency, to lower supply chain risk, and to benefit from better value and lower cost construction procurement.

In essence the Honda philosophy is one that, starting with a train of thought, says that problems start small and get bigger if they are not identified and resolved. It is perhaps easier to imagine this perspective from the production line analogy where the same problem can repeat itself hundreds, even thousands, of times a day if left unchecked.

---

2 The late Sir Michael Latham (1994) *Constructing the Team*, www.constructingexcellence.org.uk.
3 Bayfield, R. and Roberts, P. (2004) *Insights from beyond Construction: Collaboration – The Honda Experience*, www.scl.org.uk.
4 Bayfield, R. and Roberts, P. (2004) *Insights from beyond Construction: Collaboration – The Honda Experience* p 15 www.scl.org.uk.

The counter scenario of project delays and un-collaborative working was researched by Anthony Morgan and Sena Gbedemah of Price Waterhouse Coopers LLP (amongst others).[5] Their research on several government contracts cited 'lack of effective project team integration between the client, the supply chain and the suppliers' (i.e. lack of collaboration).

Another exemplar study for all the wrong reasons is that of Holyrood House, the Scottish parliamentary building. The first project estimate was for a simple refurbishment of an existing building and was reported at £10 million in the devolution white paper of September 1997. The white paper also contained an estimate for the cost of constructing a new building for the Parliament at approximately £40–50 million. This estimate was made prior to the identification of a location or a design.

In the summer of 2003, with construction nearing completion, project costs had soared to £374 million and Parliament announced that there would be an inquiry into the Holyrood building project. Lord Peter Fraser of Carmyllie QC was subsequently appointed to carry out an independent inquiry. Lord Fraser published his report in September 2004. It is a scholarly document of some 300 pages that followed 49 days of hearings and much further investigation and research. When presenting his report Lord Fraser commented as follows:

> This unique one-off building could never ever have been built for £50m and I am amazed that for so long the myth has been perpetuated it could. As it is now completed the building was bound to have cost in excess of £200m and that could or should have been anticipated at least from the time of the Spencely Report. It is difficult to be precise but something in excess of £150m has been wasted in the cost of prolongation flowing from design delays, over-optimistic programming and uncertain authority. That is an exceptionally regrettable and an uninspiring start for a new Parliament. However I trust that the exposure that my Inquiry has exhaustively given to these issues will have had the necessary cathartic effect that the Presiding Officer sought.

The Fraser report provides invaluable and publicly accessible evidence as to what can happen where there is a lack of early warning, lack of collaboration and lack of leadership.[6]

Returning to the question of 'why early warning?' The benefits of an 'early warning' project culture lie in preventing financial cost overruns to the project and/or identifying wider potential benefits to the business. In other words, the early warning approach is a proven method of both derisking and opportunity identification.

Many of those who have experience in the dispute industry will recognise that big problems were once small problems. Most disputes could have been avoided had an early intervention been possible by a key party or individual. An early warning approach enables problems to be identified and addressed before they become too large and expensive to be managed by the project team.

Returning to the question of 'why collaborate?', perhaps the question should now be phrased 'why don't you collaborate?' After all, there is a strong body of empirical

---

5 Morgan, A. and Gbedemah, S. (2010) *How Poor Project Governance Causes Delays*, – www.scl.org.uk.
6 Lord Fraser's report into Holyrood, together with the evidence, witness statements and other key information can be found at www.holyroodinquiry.org.

evidence that shows that if your project is not set up for collaborative exchanges of information, then there is a high risk of delays and overruns. The NEC4 is flying the flag for collaborative working. Those who use NEC4 will need resources and training to help them work collaboratively as no contract will achieve collaborative working on its own. The NEC and the early adopters of NEC4 deserve great credit for identifying the challenge of collaboration and sending a message to the market that this is the way forward.

14

Development Contracting – An Efficient Way to Implement Major Projects

Jon Lyle

BBuild (Hons) UNSW, Development Director, Grocon, Australia

14.1 Introduction

As a person who has worked in both contracting and property development, I will demonstrate in this chapter views from both sides of the fence, with an emphasis on project delivery. While many of the large projects discussed in this book will deal with infrastructure delivery, my experience and hence this chapter will deal with major projects within property development.

This chapter mainly deals with the implementation of the projects but the procurement phase of the project needs to follow the same guidelines.

14.2 Major Projects Are Unique

If you take any big business – Coca Cola, McDonalds, or your supermarket chains, you will find they have a standard routine, where they face many of the same challenges every day – suppliers, staff, shop leases, shoppers, etc. What works in say Sydney works basically the same way in Los Angeles, albeit with changes for the local market.

This is never the case for individual major projects as they tend to be totally unique, so while there are a standard array of challenges you then need to add the special constraints and restraints: everything from A to Z Archaeological discoveries to Zoning classifications. Major project implementation is a tough industry and a 'trap for young players'.

Contracting has changed dramatically over the last two to three decades, exacerbating the complexity as scopes have widened with more tasks such as novated design consultants, greater statutory responsibilities and even lease-up obligations being passed to the contractor.

The contractor is being increasingly linked to the property development cycle, which drags the contractors, sometimes kicking and screaming, into the even bigger and more complex property development cycle.

Whilst the contracting industry can be black and white where $2 + 2 = 4$, the property development industry would take the same equation and want to get it for three and sell it for six, in other words, it's a 'grey' area.

Global Construction Success, First Edition. Charles O'Neil.
© 2019 John Wiley & Sons Ltd. Published 2019 by John Wiley & Sons Ltd.

Tough contracting jobs can make the contractor feel like they are 'knee deep in crocodiles' but when you add the property development complexities you can tip into the same pond a few bull sharks, some piranhas, and a few sea snakes for good measure.
There are two options:

1) Attempt to 'silo' or compartmentalise the works, as done in years past, while attempting to distance the works from other facets of the property development cycle, but regrettably finding oneself being dragged in and reluctantly accepting extra loads along the way.

2) Recognise the powerful position you are in as a 50% plus cost centre within the property development cycle and use that to embed your team into the cycle. In this way you can recognise the opportunities while circumnavigating the pitfalls due to the knowledge gained by your embedment. Know what you are getting into or, as Sir Robert Baden Powell told us, 'be prepared'.

The latter option is the way to succeed. The contractor has to be prepared to deliver full developments and hence a contractor has to become a development contractor and partner the developer/investor.

14.3   Commitment and Costs

Commitment to a large project has to be complete, by all, and for the total project duration. Obviously specialist input is required at stages but the main project owner must have a continuous role based on the initial tender proposal, and the execution team should be part of the evolved negotiated contractual position.

Time and time again it can be seen with, or sometimes without, the benefit of hindsight that a project was taken on for a variety of the wrong reasons including:

- *To buy cash flow*. Short term solution to a long term problem.
- *Reinforce company profile*. Can stretch a company's resources often to breaking point.
- *Feed the associated company's development arm*. Resulting in both companies suffering or going out of business in the worst case scenario.

Projects should only be taken on to realistically achieve the project margin and inbuilt bonus provisions that can be identified by true development contractors. Any tender must be based on very thorough research by experienced teams remembering the expression, 'Hope for the best but plan for the worst'.

Creating a continuous project owner creates responsibility and that means from tender through to the DLP to the point where the project becomes an operating precinct.

The project needs to become an 'operating company' with the project director being in effect the CEO.

Set the project up for success not failure.

The project will become a 'living entity' that needs to be fed.

One critical requirement is to create a project account so that senior management manage their cash position accordingly and the only company transfer should be to and not from, with assistance to the project account only in times of difficulty or initial periods.

The project account can never be a conduit for company expenditure on everything from new sites to a director's Maserati. In the United Arab Emirates (UAE) I commenced work on an AED700 million stage of a large mixed use project with AED1 billion of presales revenue only to find that the project revenue was used to buy more development sites, leaving us with a drip feed funding scenario that was unworkable.

Project costing should be total but spurious overheads should not be considered. If the national construction manager spends 10% of their time on that project then that becomes a project cost. Don't park corporate overheads.

Done correctly the development contracting business can be very lucrative, done half-heartedly it's a highway to hell.

14.4 The Tools for Successful Development Contracting

The development contractor must understand and implement the V I P S

- **V**ision and **v**alue
- **I**dentify all stakeholders
- **P**lan, **p**rogramme, and **p**rocesses
- **S**ustainability checks.

14.4.1 Vision and Value

Don't get me wrong, we need the governments and developers to think outside the square and think BIG. History and the current day gives us examples such as

- Snowy Mountain Scheme in Australia
- London's Cross Rail
- Rendeavour's Tatu City in Kenya
- The exploits of **Sheikh Mohammed bin Rashid Al Maktoum** in Dubai.

The development contractor must use their experience to embrace, embellish, quantify, and scope out the vision on sound financial data and look for opportunities while guarding against risk.

Realism is required so that the vision can be thoroughly examined and tested with adequate market research commensurate with the scale of the development so that **everything is clearly understood and scoped** in order that the vision can then become a reality.

The UAE property market was out of control in early 2000 when inexperienced property developers were allowing architects to design their 'wet dreams' and then were surprised to see the design and construct (D&C) (D&B for UK) budgets spiralling out of control. They were only saved by the property prices spiralling higher but when 2009 came along it all came to a grinding halt when the prescription required was a large spoonful of realism.

As the world gets smaller we all find ourselves working around the globe as we move from one emerging market to the next, but the assumption that neighbouring countries are similar can be the first step to disaster. While Indonesia and Malaysia share a common language, a lot of similarities stop there and they are now very different markets.

Even within countries such as the UAE there are individual emirates such as Abu Dhabi and Dubai that have completely different statutory regulations when it comes to building codes – know your area.

Invest in local teams – co or joint venture, or peer review, but come up with a formula that starts with local knowledge that can be complemented with expatriate expertise.

Nowhere is like 'home':

- Be cognisant of local conditions and production outputs.
  As a simple example, the calculation for rebar – in Australia planners would use 750 kg/person/day but in the UAE production rates are as low as 125 kg/person/day or 17%.
- Keep your standards.
  On an Abu Dhabi high rise we were offered a large saving by using spandrel panels that contained combustible material, which we thankfully refused given recent fire tragedies.

You need to REALise and LOCALise to properly value the vision, remembering any decision has to be based on three disciplines – commercial terms, design constraints and constructability.

Wherever possible, look at Value Add possibilities knowing that margin comes from two areas, 'revenue up' and 'cost down', remembering the former can be far more lucrative than the latter.

The industry is moving to predominantly D&C so whether preferable or not a development contractor must understand and utilise the design process.

Within this process there are three ever growing key risk areas:

- *Design complexity*. Developers continue to push the design envelopes to 'unbuildable' concepts, which effectively means 'unpriceable' projects.
  On an UAE 80-level high rise office building the designer included sloping structural blades from GL to a three level transfer floor L19, 20, and 21 where the main structural core commenced. The design was further complicated by a structural diagrid facade, meaning the facade was on the structural critical path. Design rationalisation reintroduced a central structural core, leaving the sloping blade walls as Architectural elements and deleted the diagrid but retained the look by aluminium sections. This represented an AED600 million saving from the construction cost, made the project buildable but most importantly made the development viable.
- *Design risk and responsibility*. While the concept of D&C is based on optimising the design by the teams that have to construct the works based on performance criteria, the principle of D&C has been mismanaged to an unfair or inequitable position. Developers are designing the works through concept and schematic design and into design development stages just stopping short of IFC/AFC documentation and then insisting on design warranties. It's just not practical.

  It is interesting to note that in a legal case in Australia it was ruled that the developer was guilty of design exhaustion and was not able to contract out of the design responsibility, whereupon it became the subject of a court case.

  This does not mean there is no opportunity. In Kuala Lumpur our structural design was specifically designed to suit the locally available systemised jump form systems in the local market.

- *Design omission.* Either from deliberate intent or blissful ignorance developers do not define true scope(s) and combined with the continuous effort to reduce costs developers will reduce design costs to a point where it is difficult for the design consultants to document true scopes. When design is lacking or unclear, the development contractor must reverse brief so that you quantify and qualify before an item can become ambiguous or a dispute.

The message here is to evaluate all of the above well and truly and make proper allowances or in rugby terms at a loose maul 'use it or lose it'.

Design management, once the task of the client, is rapidly becoming a development contractor's responsibility so you need to use it to your gain.

While above we referenced partnering the developer, the development contractor needs to enhance their position by understanding the development thinking with respect to vision and value.

On an office tower in Jakarta in the early 1990s the central core design was based on 700 mm thick core walls, which, when value engineered by our international peer review engineers, were reduced to 350 mm. Notwithstanding the savings on concrete, the additional nett saleable area when capitalised was worth an additional US$5 million.

In another residential development in Sydney the specification called for column free space for the convenience retail below the residential tower. While column free space is a requirement for A grade commercial office space it is hardly a major requirement for convenience retail. The introduction of well-planned columns saved AU$1.5 million, which is a considerable sum from a AU$50 million build.

There is a twofold message here

1) Using a food analogy. 'If it doesn't smell right it's probably off' – challenge the unrealistic.
2) Recognise opportunities, in the Jakarta example above would you prefer to take the US$300 000 concrete savings or be part of the US$5 million uplift.

Never get swept up in the vision to not test; test and test again for value rationalisation. Create development models to run the sensitivity analyses and 'what if' scenarios.

REALise and LOCALise the vision and true value to make your project plan robust to protect margin, seek out opportunity and guard against risk.

14.4.2 Identify All the Stakeholders

Identify all stakeholders on day 1 that 'may play' a role in the development cycle. A stakeholder in general terms is either an entity or an event that may contribute or hinder a project. Stakeholders come in many 'shapes and sizes' but can be split into eight main categories:

1) Company
2) Funders/financiers
3) Partners
4) Statutory authorities (inclusive of associated parties)
5) Consultants
6) Construction
7) Sales and marketing
8) Operators.

This identification and management reinforces the need for continuous management ownership as previously referenced.

This needs to be exhaustive so that when, say, an obscure environmental group raises its hand in an attempt to push its objectives against the project, you are ready; this stakeholder/risk will have been previously tabled and hence you have a strategy (albeit in draft) in place.

Whether preferred or not the contracting environment links all stakeholders in so many ways, from backing down of serious developer obligations to tripartite agreements with unconnected third parties and a series of other 'connections'.

We illustrate in Figure 14.1 the stakeholder communication globe.

Each and every stakeholder will have specific perspectives, goals and objectives and the 'communication' lines crossing points indicate or highlight the need for clear management strategy implementation to ensure overall development or project goal achievement.

For each and every stakeholder you need to assess the time and deliverable scopes for engagement and possible contract so that their involvement is timely and effective and you are the controller that engages and disengages. Treat this like a radar screen and consider yourself an air traffic controller. Forewarned is forearmed.

Stakeholders can be synonymous with both opportunity and risk and these too need to be identified for your risk and opportunity register (R&O), which needs to be updated monthly. To enable all these to be included we need to schedule them like cash flow, allocating a clear period to identify to senior management who has to own, understand, and monitor all R&O to ensure that a risk, using medical analogies, cannot bleed out and it is cauterised immediately. Too often we see a band aid solution for something that needs more major surgery. Time and time again we see an unplanned situation arise

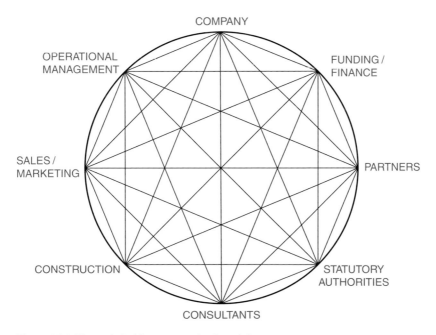

Figure 14.1 The stakeholder communication globe.

and rather than address it immediately and take the hit it is allowed to hang out there unresolved, which can mean a cost of 1× grows into 10× and a Y month time loss.

Where a stakeholder represents a risk, this risk must be cash flowed like revenue or cost so that it is clearly identified in the total project but allocated to a set period, so a risk that may eventuate in say Q3 of Y2 of a project may not require senior management input until Q1 of Y2.

14.4.3   Plan, Programme, and Process

Deliver on time and on budget and exit cleanly.

As projects become larger the implementation of sensible development staging and individual project planning is paramount. This includes the introduction of flexibility to allow change of use or even an exit strategy given a changing market. Careful consideration is needed to maintain economies of scale without over extending or increasing project risk. Once the overall development staging is identified, the team needs to drill down on individual project staging and in micro detail when required so that set deliverables and processes for all stakeholders can be specified at each and every one of the standard project stages.

Table 14.1 is an example of a development strategy matrix from a residential development that can be populated with the stakeholder deliverables. It would be preferable for the development contractor to begin involvement around stage four or five, Concept Design & Options.

Table 14.1 stages are seen as three monthly intervals but preferably intermediate monthly milestones should be incorporated as 'ready reckoners' to the project plan's accuracy.

It is essential for corporate governance and to ensure project progression is based on solid foundations. It is also essential for communication and education/training of the team.

There are many examples that the prudent development contractor should know so as to ensure the development.

The contractor and developer/investor relationship is complementary.

For instance on a residential project the interconnection between pre-sales, design progression and funding release is of paramount importance. It should be acknowledged that the end of the schematic design requires all areas, structural elements, and plant rooms and risers to be locked down, which means then, and only then, can you allow the marketing teams to compile their brochures, confident that there will be no misrepresentation. The development contractor should then know what percentage of pre-sales is required by the funding facility to release construction funds. While a contractor may say that's a developer's problem, the good development contractor knows 'no money, no honey' and will assist in getting the funding mobilised to ensure he will be paid.

As another example, each stage of the construction staging should have its own list of deliverables, and referring back to Table 14.1, taking stage 2 of construction or stage 14 of the entire cycle, again a definitive list of deliverables must be compiled and then the QA/QC processes kick in to create the 'hold' point for review and sign off. In Table 14.2 we have listed typical deliverables for this stage on a residential project, which then become the reporting requirements that the project owner should sign off before agreeing that the project can proceed to the next stage.

Table 14.1 Stakeholder strategy, responsibilities and deliverables matrix.

Project Stages	Planning						Design						Construction					Operation		
Strategy (Greek "στρατηγία"— stratēgia, "art of troop leader; office of general, command, generalship" is a high level plan to achieve one or more goals under conditions of uncertainty. Strategy is a) important because the resources available to achieve these goals are usually limited and b) about attaining and maintaining a position of advantage over adversaries through the successive exploitation of known or emergent possibilities rather than committing to any specific fixed plan designed at the outset.	1	2	3	4	5	6	7	8	9	10	11	12	13	14	15	16	17	18	19	20
	Identification	Acquisition	Briefing	Concept Design	Options	Statutory Initial	Schematic Design	Statutory Approvals	Marketing	Design Development	Tender / Early Works	Final Design - IFC	Const. Project Est.	Contractor Design	Const. Monitoring	Const. Finalisation	Const. Completion	Const. Handover	Operation - Initial	Long term Mgt.
STAKEHOLDERS																				
Company																				
Inter company																				
Internal divisions																				
Strategic Alliance partners																				
Funders/Financiers																				
Equity partners/Shareholders																				
Main debt funding																				
Mezzanine funding																				
Client Milestone funding																				
Escrow trustee																				
Financial consultants																				
Partners																				
Master Developer																				
Sub Developers																				
Joint Venture Partner(s)																				
Historical																				
Current																				
Future																				
Adjoining developers																				
Adjoining Owners																				
Competitors																				
Political Entities																				
Public Entities																				
Community																				
Environmental																				
Heritage/Archeological																				
Statutory Authorities																				
National Plans																				
Townplanning																				
Approval (Licences)																				
Approval (Technical/Design)																				
Infrastructure - Supply/Utilities																				
Infrastructure - Civil & Transport																				
Infrastructure - Social																				
Titling																				
Ancillary Authorities																				
Reporting																				
Testing																				
Financial Certifications																				
Approved 3rd party consultants																				
Private Utilities																				
Consultants																				
Financial & Legal																				
Management																				
Technical/Design																				
Peer Review / VE consulatnts																				
Survey																				
Marketing/Sales																				
Construction																				
Contractors																				
Infrastructure																				
Enabling Contractors																				
Main/Principal Contractors																				
Sub Contractors																				
Separate Contractors																				
Nominated Contractors																				
Suppliers																				
Sales & Marketing																				
Agents																				
Customers																				
Lessees																				
Contracted Clients																				
Operational Management																				
Building/Facilities Management																				
Strata Management																				
Property Management																				
Operators																				
Licencees																				

Table 14.2 Standard report including.

Marketing/departures audit
Statutory audit
Consent and CC status
Contributions status
Consent register update
Standard area schedule GFA/NSA
General letting status
Financial summary forecast
Consultant certification
Programming
Contract vs target
Milestone reporting
Provide as built boundary survey
Re-measure strata plans off AFC's
Review all marketing plans for any changes
Review all parking and storage allocation areas, access, levels
Overview total 'metering' diagram to delineate areas
Shop drawings review and certification
Fire engineering requirements, if applicable
Room data sheets sign off
RCPs – report any areas less than 2700 or 2400
Mechanical
Bulkheads, ducts, FCU and condensor locations
Provision for future AC
Location of plant outside lot
Risers, ducts running in other lots

(Continued)

Table 14.2 (Continued)

Typical floor plant
Roof plant
Acoustic and aesthetic checks
Exhaust and louvre locations
Access provisions to all plant
Electrical
Metering diagram
Standard point layout – GPO, MATV, intercom, switching (internal and external), lights
Kiosk substation location and access
Security system
Common lighting – internal, external, building
Hydraulic/plumbing
Notice of requirements and S73 update
Extent of gas and water (internal and external)
Drainage routes in basements
Fire
Access and locations
Functionality
'Walk throughs' to establish additional balcony, terraces, storage, standard common areas, clearances
Privacy
Review all balcony areas for screens
Finishes – 90% sign off
Presentation by area
Samples
All apartment schemes
Schedule any approvals required up to next stage

Planning and programming go together, and while projects may be able to be programmed with the likes of Prima Vera or Microsoft Project, they cannot be planned as this requires 'big picture' thinking and very often the ability to work backwards from a specified or target completion date.

Programmers are like computers as their output is only as good as the information inputted. There is no substitute to getting the team around the whiteboard to draw the project plan, working out the major milestones, conditions precedent, and real trigger points.

How many times have you been handed a 60 page A3 computer program that will take hours to even understand let alone logic check. It's therefore far more transparent to create a graphical project plan like the one in Figure 14.2 that snapshots all the big picture information onto one sheet, from site plans, crane radius diagrams and of course the programme activities linked.

Scale will not permit but the project plan in Figure 14.2 is from a twin office hotel project in Kuala Lumpur where the developer had to open the hotel prior to the 1998 Commonwealth games so the required time frame and local ground conditions dictated a top down construction sequence (5 basements with 42 m barrettes) with a jump steel start from B1 to L3 to allow construction commencement of the typical hotel floors while the basements were being excavated. All of that is represented below.

As a communication tool these project plans are second to none.

After the plan is established then detailed subprogrammes using proprietary computer programs can 'fill in the gaps' and then the tracking and monitoring systems must cut in, as discussed previously.

There are many other tracking tools including a simple concrete graph compiled by working out monthly concrete pour requirements to meet programme and then any manager can track the job by receiving a cumulative concrete pour tally each week. Concrete waits for no one and never lies!

The bottom line is **there is no excuse not to diligently track projects from any distance** to ensure that when the 'train veers off the track' you as the project owner are there to nudge it back on line.

14.4.4 Sustainability

While some will jump to visions of wind turbines and roof structures awash in solar panels the truth is that the most important sustainability item is that the project is a success – financially or in target use (e.g. civic requirement).

The project must achieve its goals so its lifecycle is met or exceeded because an unsuccessful project that doesn't complete or doesn't sell means that it will create negative brand equity for all concerned, and the drain on environmental resources to make good or rebuild a new development is far outweighed by minor environmental design(s) within the original development.

While it may not be a development contractor's obligation or responsibility the prudent development contractor needs to understand how the development achieves sustainability and hence financial success. It then becomes incumbent on them to assist this process with any conditions precedent or subsequent to funding agreements. Understand and work the system.

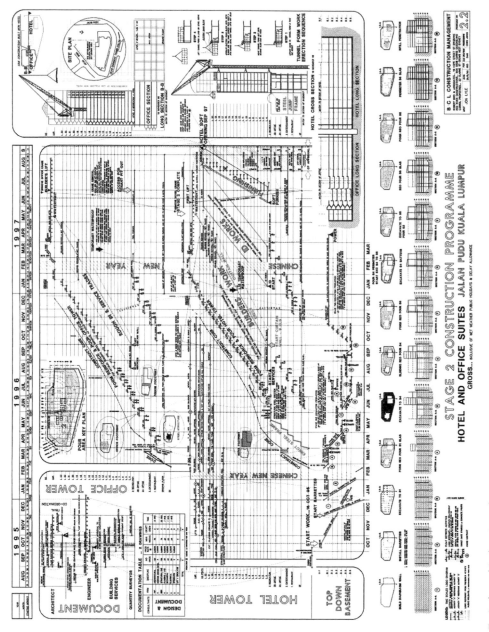

Figure 14.2 Typical graphical project plan.

Once you understand this you can act and contract accordingly with more of an assurance that the development is 'sustainable' and the client has or is getting the cash, as in everything 'cash is king'.

14.5   Conclusion

What does this say to contractors implementing major projects – you need to become 'development' contractors, fully immersing in the entire property development cycle using the VIPS to understand all the areas that can help and hinder your journey so that it becomes a walk in the park that others, unprepared, would see as a possible minefield.

A successful development contractor will partner the developer/investor and then will be able to deliver the following oath:

We came,
We saw,
We concreted and then,
We did LUNCH.

15

A Critical Review of PPPs and Recommendations for Improvement

15.1 Introduction

Public Private Partnerships (PPPs) are widely used around the world today as an alternative means of providing infrastructure. They involve the use of construction, operating and funding skills from the private sector, whilst simultaneously reducing the overall project risks borne by the public sector (governments), through transferring procurement and operating risks to the party that is *'best positioned to manage the risks'*. PPPs are sometimes known by other names, such as Design, Build, Finance, Operate (DBFO), Private Finance Initiative (PFI) and many others. It is regrettably a form of procurement that has developed its own confusing mix of acronyms and jargon! For this chapter we shall stick to the term PPP, but this encompasses all of the various similar forms of public–private procurement.

It is often overlooked that private involvement in the procurement, operating and finance of infrastructure is nothing new – it has been that way for centuries, and PPP (or DBFO, or PFI) are merely the current names for something that has been long established around the world. The great infrastructure expansions of the nineteenth and twentieth centuries were significantly funded, delivered, and operated by private sector initiatives – think of the railroads of the US, the early underground railways of London and many other cities, and the great steamship building programmes of the 1800s.

Historically these projects have always arisen as part of an associated expansion in economic trade or political influence. This is an important point. The interests of the private sector in this form of procurement throughout history have largely been motivated by commercial imperatives and not by altruistic ideals. This has often led to tensions between the private sector's desire to produce commercial return (through the efficient delivery and management of infrastructure) and the primary objectives of the public sector (being the improvement of infrastructure and services for community benefit – but with the government not always having the funding or delivery capacities to do this alone). These diametrically opposed objectives remain today at the core of much continuous debate about the real value of procurement by PPP methods. 'Successful' PPPs are characterised by the parties to any project recognising these differences, but also by the value of the third 'p' (partnership). Although the parties may not have fully aligned objectives, a PPP project can be a successful outcome for all parties if they are prepared to work in a collaborative and understanding model of partnership.

Global Construction Success, First Edition. Charles O'Neil.
© 2019 John Wiley & Sons Ltd. Published 2019 by John Wiley & Sons Ltd.

Not all PPP projects are successful. This is nothing new. Referring back to our earlier examples of US rail expansion, underground systems and steamship building – these areas are full of examples of failed projects. We laud the great engineers of the past, but sometimes forget the calamitous nature of the private sector funded infrastructure projects in which they were involved. Society has benefited historically from two aspects of such private sector involvement in the provision of infrastructure – without private involvement many of the projects would never have been commenced, governments being unwilling or unable alone to bear the funding or risk of the early development stages. And when the private sector has participated and the projects have failed as investments, it has generally then been the private sector that has been left to pick up the pieces and restructure, certainly at considerably less risk and cost to the public sector than would have been incurred if the projects had followed procurement and delivery solely by traditional public sector methods.

Nowhere is the above more recently demonstrated than in the UK with the insolvency of Carillion, a major private sector provider of construction and operating services to PPP projects. Ultimately the responsibility and cost to replace most of the failed Carillion services and to ensure their continuity has fallen directly on the private sector equity owners of the PPP project companies. This is risk transfer in direct action, and this key message is often lost in the political and media storms that such occurrences attract. In the case of Carillion, the majority of public services continued to be delivered with minimal interruption after the insolvency, and the additional costs of maintaining these services and replacing the failed provider were borne directly by the private sector. The worst outcome of Carillion's collapse has been that two of their hospital projects in the U.K. have had to be terminated as PPPs during construction (as of September 2018), which will incur considerable additional cost to the taxpayer and cause significant delay in the provision of the hospitals. This should be regarded as a result of the appalling mis-management of the company, not as a breakdown in the PPP process, because there has been a very high completion success rate with more than 700 PPPs in the U.K. to date.

We will see in this chapter that PPP failures occur for a variety of reasons. These can be broadly defined into three (interrelated) categories:

- Economics. The project is financially unsustainable. For example, the new toll road that assumes revenues of 40 000 paying vehicles per day, but in reality attracts only 20 000. This happens too often, and is not always the fault of the traffic forecasters – there can be inherent political and commercial incentives for exaggeration of the likely usage in order to ensure that a particular project actually proceeds, irrespective of its overall viability. Many failed PPP projects should never have been procured by this method, but were driven into it by political or economic considerations unrelated to the actual suitability of the PPP as a delivery method.
- Partnership. Either or both the public and private sector participants in a particular project fail to understand the needs and objectives of the other party. Long term partnership projects require collaboration and understanding, a win–win mentality.
- Initial definition. Failure to understand at the start of procurement what is actually required and to then express this in a clear manner has caused many PPPs to be regarded as 'unsuccessful'. A downside of the PPP procurement method is that it works best with very clear specification and design at the start, and does not respond well to continuous unplanned changes in specification throughout the delivery and operating stages. This is not to say that change cannot be incorporated – where change is anticipated and a mechanism is agreed at the start to manage this aspect through

the duration of the contract, then PPP contracts can be very successful. However, a failing to think and plan for the future can be very expensive!

These reasons for failure are investigated more fully within subsequent sections of this chapter.

In recent years, the use of PPPs has decreased in some regions, such as Western Europe. However, it remains very strong and is growing in many other regions, particularly in China, India, North and South America, and Africa. It is notable, at the time of this book being published, that China actually has the largest pipeline of PPPs of all countries.

As PPPs comprise such a significant sector in the global construction industry, and because the jargon around this procurement method is often so complex and confusing, it is appropriate that we devote considerable space and detail to the sector within this book. The challenges and benefits of PPP procurement are not difficult to understand, and we seek to explain these within our comprehensive review.

Volumes of material have been published on PPPs since the early 1990s, covering the concept, financing, contracts, implementation, operations and maintenance, etc. Some of it is quite theoretical and academic, but much of it is also good practical information and guidance.

In this chapter we will be concentrating on:

- The merits and otherwise of the PPP process (the pros and cons)
- How PPPs can and should be improved
- The partnering and the human factors involved, which are critical to the success or failure of PPPs.

There are two particularly contentious areas that we will address, followed by practical hands-on summaries of the key areas of implementation,

- The ongoing ideological and economic debate – yes, much has been written on this but we want to revisit the pros and cons of the concept.
- This is because the PPP concept does show some flaws and can benefit from review and possible updating. Apart from improved contractual formats, the basic structure and implementation of PPPs and their transfer of risk has seen little change for many years. As a result, many potential private sector investors, construction contractors and facilities management (FM) service companies are now shying away from PPPs because of the flaws in the process and the losses that they have sustained through bad experiences.

In the following sections we address the key aspects of PPPs that require the significant human behavioural input that is crucial to success and, in particular, how this relates to successful 'partnership' within the concept of PPP projects.

Note: the author Charles O'Neil has had the following recent involvement with Chinese infrastructure development.

'2014 – travelled 5000 km from Kashgar in the far west to Shanghai, witnessing the extraordinary development of infrastructures en route, including motorways and high speed rail, steel mills, power stations, wind turbine and solar farms, large scale industrial parks and new airports and cities such as Urumqi.
2015 – appointed to an international subcommittee to review the Beijing Arbitration Commission (BAC) Dispute Board Rules.

2016 – panellist at the BAC London Summit and speaker on '*Implementing PPPs success-fully – including areas of risk and potential for dispute or failure*'.
2017 – guest lecturer lecture on PPPs to a delegation of senior Chinese bankers at Oxford University, UK.'

15.2 Proponents and Opponents

Let us start by considering how a typical PPP works. The public sector will contract with a special purpose company (SPC) from the private sector. The SPC will be responsible for designing, building, operating, maintaining and financing a piece of infrastructure for a fixed period (typically 25–30 years). The public sector will pay the SPC a fixed regular fee for the provision of the infrastructure. If the service is not provided to the correct standard, then the fee is reduced. At the end of the concession period, the SPC has to return the asset to the public sector in a fixed condition. The SPC will engage directly with builders, operators and financiers to deliver the project. The SPC is a 'stand-alone' entity. It has shareholders who provide equity to the SPC, but any debt funding it raises to support the project must be done on the basis of only that project's economic strengths. The financiers have no direct recourse to the shareholders of the SPC beyond their equity invested in the SPC. As a result of this constraint, the funders have appropriate strict controls in how the SPC is managed and run.

A typical PPP (social buildings – hospitals, schools, etc.) will generally comprise share-holder equity in the SPC of around 10% of the total funding requirements of the project. The required equity percentage increases as the asset risk type increases (for example a toll road PPP may have a 30–40% equity level, because the revenues to repay the debt are directly dependant on the traffic tolls received, and therefore are more risky than a social building PPP where the revenue is a secured by direct payment from the public sector, subject only to deductions arising from poor service provision).

In general terms, the supporters and opponents of PPP can be summarised as follows.

15.2.1 Supporters

- Governments and Authorities who acknowledge PPPs as being a viable alternative funding method to provide and maintain infrastructure in a long-term and efficient manner, whilst at the same time significantly reducing the direct public sector debt burden and the element of project risk assumed by the public sector in the provision of infrastructure. (Note that these can sometimes be '*reluctant supporters*' – PPP in some circumstances may be the *only* way in which governments and authorities can readily procure new infrastructure in circumstances where their internal funding, economic constraints or risk transfer limits would prohibit them from direct public procurement).
- Providers. This category includes equity investors, building contractors, suppliers, subcontractors and facility management service companies with long-term operation and maintenance (O&M) contracts.

- End users. These are the wide range of public and private entities that benefit directly from delivery under the PPP model. Examples would include:
  - Hospital trusts or education bodies, including their employees and patients/students who benefit directly from utilisation of the new facilities (which generally replace old, inefficient and poorly maintained infrastructure).
  - Commercial businesses that might benefit economically from new, more efficient transport infrastructure.
  - Governments and authorities that may benefit indirectly from general economic growth in their region as a result of providing and maintaining attractive new infrastructure.

15.2.2 Opponents

- Ideological individuals and bodies, usually vocal in the media or at conferences (primarily political) who are fundamentally opposed to any involvement of the private sector in the provision of public infrastructure. They argue that infrastructure could and should be provided directly by the public sector in a more economic or efficient manner.
- Disgruntled providers. This category includes an increasing number of equity investors, building contractors and O&M companies who are no longer participating in PPP markets. Their reasons include that they have had bad experiences in the past, or because they find the current format and risks commercially unacceptable. A particular issue is the relatively high up-front bidding costs that these bodies have to bear, at their own risk, prior to any project award.
- Disgruntled end users. This comprises a relatively small group, usually frustrated with aspects of how the PPP works in practice, particularly with how even simple scope changes can become very complex and expensive during the operating period of the assets. This is a justified criticism of some PPP projects where the framework, scope and change mechanisms have been poorly conceived and drafted at the start. Given increasing overall budgetary pressures, some bodies are becoming uncomfortable with the long-term and fixed financial commitment required for a PPP project. Many state bodies find that this restricts their flexibility for their future overall budget allocation.

15.2.3 Addressing the General Pros and Cons of PPPs

Around the world there are many government and research studies that highlight the inefficiencies of infrastructure procurement by the traditional 'direct' route. Typically, these studies find that projects are delivered with delays of 20–40% to the original time schedule and with a typical similar increase against the planned budget.

A primary benefit of PPPs is that they ***transfer the risk of timely and cost effective delivery from the public to the private sector.*** If a project is delivered late or costs more than planned to deliver then, under the PPP model, that risk is the responsibility and at the cost of the private sector provider and is not an additional burden on public finances. Analysis of completed PPP projects demonstrates that these are significantly better at providing on-time or early delivery of a new asset into operation when compared with direct public sector procurement methods.

A further major benefit of the PPP method is that it ***provides for a fixed standard of service and maintenance for a fixed fee to the public sector for the entire long term duration of the contract***. In addition, at the end of the defined contract period under a PPP contract, the private sector is usually obliged to ***hand the asset back to the public sector in a defined quality of condition***.

A common argument against PPP procurement is that it would be ***cheaper for the public sector to procure directly*** than to engage the private sector to deliver and manage (and to charge a profit for doing so). At the start of a project it is quite normal for the public sector body to produce a comparison between the overall costs of traditional procurement versus the costs of procuring and maintaining the same asset under a PPP model. This is often termed as the public sector comparator (PSC). The PSC has turned out to be one of the most misused and manipulated data sources by both the supporters and the opponents of PPP.

- Supporters of PPPs have undoubtedly in the past 'constructed' PSC models specifically to show that the PPP procurement method is financially more efficient than traditional procurement. This has often been done as part of the final push to get a PPP project over its final approval lines.
- Opponents of PPPs consistently underplay the PSC cost benefits of a PPP, particularly with regard to the real costs of long term maintenance provision. Historically, in the UK public sector bodies had a very poor record of funding the maintenance of their infrastructure – a direct consequence of this became the need in the 1980s to establish a major PFI/PPP procurement programme in this country. It had a large stock of school and hospital infrastructure, built in the 1950s and 1960s, that had been very poorly maintained and were therefore all reaching an unserviceable state at the same time. The Government did not have the direct financial capacity to fund replacement, and therefore the need for private sector involvement became essential.

It is often argued that the ***public sector can usually secure long term debt funding at cheaper rates*** than the private sector. Generally this is true. However, public sector bodies are constrained by regulation on the total amount of debt funding that they can source, and PPPs can therefore support the public sector as part of a balanced portfolio of debt sourcing for long term infrastructure provision. It should also be noted that, in these times of historically low interest rates, the margin difference between private and publicly procured interest rates is relatively small.

Traditionally, PPP debt funding has been structured on a standard debt/equity model (say 90/10 for social infrastructure projects). However, in recent years Canada has initiated the structuring of projects in which the state partner has injected a substantial sum as a capital contribution, in some cases 50% or more of the total capital cost over the life of the concession as determined in the financial model. The private sector still assumes most of the operational risk and the risk of return on their investment equity, but with the state having substantial 'skin in the game' the two parties really do become a closer partnership with more aligned interests. This funding method has now become increasingly popular around the world.

In recent years, some critics have complained that private sector ***debt funding of PPP projects is not shown as part of the public sector balance sheet*** and that this therefore gives a false picture of the long term debt obligations held by the public sector. Indeed, it is claimed that such a scenario is dangerous, as it underestimates the true debt burden carried by the public sector. There is no doubt that governments (notably the UK, but also others) have benefited from being able to 'ignore' PPP-related debt and thereby show their overall indebtedness at lower levels. However, there is increasing consensus (driven partly by the changes to international accounting standards to reflect the new situation) that the exclusion of PPP related debt from overall calculations is inappropriate. State governments in Australia (notably Victoria) have shown a lead on this matter, which others (including the UK) are now adopting. Therefore, the balance sheet argument against PPP debt funding is largely receding.

Comment on PPPs by The Right Honourable Dr Gordon Brown FRSE, Prime Minister of the United Kingdom 2007–2010

There was one other set of initiatives vital for investment in our economy and public services – public-private partnerships or PPPs. For some, PPPs were simply a halfway house to privatisation. The public sector, they said, handed over control of projects to private contractors who, in return for paying upfront capital costs, charged exorbitant interest rates over the next thirty years. For me, PPPs looked different: they were a way of mobilising private funds for public purposes and offered a better route to building our infrastructure than the old ways of financing. After all, we had always hired private builders to construct our schools, hospitals, roads and public builldings. But we had enjoyed little control – leading to run-on costs, delays and often poor quality. With PPPs we were not only mobilising private sector skills for public purposes, ending the sterile and self-defeating battle for territory between public and private sectors, but building in tougher contractual requirements that secured far better value for money for the taxpayer. There were to be teething problems – some projects were poorly structured – and in an environment of low interest rates, legitimate issues about the rates of interest being charged. But, at their best, PPPs married the long-term thinking and ethos of the public sector with the managerial skills of the private; and, in truth, we could not have commissioned more than a hundred hospitals and built, rebuilt or refurbished 4,000 schools by 2010 if we had not.

15.3 Project Viability and Necessary Due Diligence

From the general discussion of the pros and cons of PPPs in Section 15.2, we now turn to address specific key issues at the level of individual projects.

In an ideal world every PPP project would be commercially viable in its own right, with user-pay revenues covering all of the construction and operating costs, the return on investment, and capital repayment. In reality there are very few PPP infrastructure projects where this happens because most are for the provision of necessary community facilities (schools, hospitals, roads etc.) that otherwise would be funded directly out of tax revenues. Therefore, public sector authorities considering entering into PPPs have certain core criteria by which they must assess each PPP project:

- There must be a definite need for the infrastructure that warrants the use of taxpayers money to fund it;
- The burden on public sector body finances of having a committed funding obligation throughout the full duration of the project must be understood.

- The PPP proposal must provide value for money in real terms, compared to the PSC (see Section 15.2 above).
- The ability of the public sector to accurately define its project requirements and scope at the outset must be achievable.
- The benefits of transferring the main risks, including the fixed construction cost, the fixed programme, and the fixed operating and maintenance costs must be defined.
- The expertise of the parties to deliver and operate the PPP project must be proven.

15.3.1   PPP Project Due Diligence

Before committing to participate in any potential PPP project, each of the stakeholders should undertake the following typical due diligence:

- *Macroeconomic*:
  - Is the facility really needed?
  - Is it value for taxpayer's money?
  - Is the legal framework acceptable:
  - Are the political risks acceptable?
- *Commercial* due diligence:
  - Assess the operational risks of the underlying asset
  - PPP concessionaire risk – credit rating/proof that they are financially solid
  - Is the bankruptcy risk remote?
  - Are the subcontractor credit risk and security packages acceptable?
  - Check the interest rate risk, if long term finance is unavailable
  - Check the refinancing risk
  - Check the exposure to inflation
- *Financial, accounting and tax*. Risks for company finances
  - Do cash flows match the forecasts in the financial model?
  - Does the FM include cash flow simulations for various scenarios?
  - Does the FM include downside sensitivities of adverse events?
  - Review the impact of downsides on key ratios, funds distribution, etc.
  - Review the tax and accounting structure – assumptions
  - Third party advisors often do the FM audit – tax and accounting report
- *Technical due diligence by industry experts:*
  - Review the technical risk profile and
  - Identify the potential upside/downside risks specific to the asset
- *Legal due diligence:*
  - Assess contractual exposure and other legal risks to asset
  - Ensure key risks are transferred to the subcontractors
  - Conduct a GAP analysis on the remaining risks
- *Insurance due diligence report:*
  - Are insurance strategies and policies in place for the asset?

The above due diligence requirements are not exclusive. Effective due diligence requires close coordination between the various categories of assessment.

15.4 Some Current Perspectives on the PPP Process

There has been a noticeable increase in private sector dissatisfaction around the world with a number of aspects of the current PPP processes. Many potential equity partners, contractors, and service companies are consciously avoiding the sector for fear of having to enter into long term contracts with public sector parties where their assumed risk profile is increasingly onerous and does not justify the level of risk that the private sector parties are being asked to assume under these contracts. There is a feeling that, in some instances and jurisdictions, PPP procurement is no longer being delivered on a 'level playing field' or in true partnership.

Failed PPPs represent only a very small percentage of the total in developed countries, but unfortunately a much larger percentage in developing nations. These failures attract a lot of publicity, which the opponents of PPPs are quick to target and highlight, but these examples also point to the need in some areas for an urgent revision of the PPP format and implementation methods. **Perhaps it is time to radically review and restructure the contractual formats?**

To contribute towards current thought and discussion, we have surveyed experienced PPP practitioners in North America, UK, Europe and Australia and have asked them to nominate key reforms that they believe could radically improve the PPP process. We quote these verbatim below.

Although a small sample, the following comments are indicative of wider and serious concerns within the industry and they are supported by what can be read in the industry press on an increasingly regular basis.

15.4.1 North America

- *Increase stipends (Bid cost reimbursement). The last rail Bid we worked on cost the developer $2 000 000 and the contractor about $15 000 000. The stipend was approximately $2 000 000 in total.*
- *Decrease the level of detail prepared in the reference design (clients design) and increase the opportunity to innovate by well-thought performance specifications from the client.*
- *Select proponents at an earlier stage based on qualifications and indicative design and budgets. Then work with the selected proponent to refine the design and a budget that would then form the basis of the committed price and corresponding schedule.*

- *The holy grail of PPPs in Canada is to be able to apply the model to smaller projects (less than circa $50 million) of which there are potentially many. Canada is a big place with lots of small municipalities (and remote indigenous communities) that have need of smaller value projects but do not have the resources or experience to develop and manage them. PPP developers prefer higher value projects because pursuit costs are significant whether the prize is a $10 million investment or a $200 million investment. Part of the problem is that a standardised contract form has not been developed that can be used 'off the shelf' by clients for smaller projects. We have not managed to get anything universally accepted in the PPP market. If this could be developed, and was* *familiar to a wide range of public sector clients, lenders and local contractors, it could unlock a much broader swathe of capital projects to the benefits of PPP.*

15.4.2   United Kingdom

- *The Construction industry has to grow up and bid more realistically – they need to be more pragmatic and every bit as tough as the client side, just because some idiots in the market will do things for a price doesn't mean that it's the right price and certainly when some consortia or company undercuts everyone else it must not be seen as setting a precedent which some major consultancies take on board when advising on client cost planning; more bravery needed by the construction industry with Board support to be able to say thanks but not for us at those prices.*
- *Require public sector equity participation at 20% and commitment to participation – to encourage greater partnership working.*
- *Private sector open book accounting from day-1.*
- *Public sector option to buy out "senior debt" without crippling break costs post construction by use of Gov't gilts with the same contractual constraints as senior debt. Emphasise this is not an equity buy-out option.*
- *Re public sector shareholding: currently going the wrong way from my point of view with public sector equity participation. We have the public sector on three projects as co/shareholder and the value add is very limited.*

15.4.3   General Points

- *Efficient and clear set out of procurement process. Procurement processes from start to final bidder selection should not take longer than 9–12 months.*
- *Professional advisors to be selected by public sector based primarily on experience and less so on price.*
- *Realistic risk transfer to private sector – we currently see in the UK the tendency that a meaningful number of contractors (if not the majority) are increasingly worried about the risk transfer during construction. Many have experienced losses and with the demise of Carillion and consequential loss announcements of Galliford Try and Balfour Beatty, the situation has deteriorated. I know of some others who will no longer play in the PPP sector.*
- *They have to be attractive (they are not in the UK) so the public element has to stop interfering with the private sector financing and focus on what it wants to achieve.*
- *Bid costs are a massive issue and need to be improved. I don't think standardising documents is necessarily the way. Perhaps a better pre-qualification process which gets you to preferred bidder based on quality and capacity, not really on the specifics of the project.*
- *Schemes need to be carefully prepared. I do not say more completely because I think for PPPs in particular you need to have plenty of flex to gain the input from the private sector. I think this becomes even more important when you look at how fast the use of buildings and infrastructure is changing at the moment. 30 years ago only 1 of the top 10 PPP companies in the world even existed so no one can imagine what we will need by the end of a standard 25 year concession period, so we need to start designing and contracting PPPs for much better flexibility, e.g. hospitals where you assume new tech will be available and it isn't, or where you assume a standard use which is simply obsolete within five years because requirements have moved on.*

15.4.4 Europe

- *Improvement of the quality and consistency of the (in particular, technical) tender documents; there is a tendency in Germany to shift more and more risks onto the private contractors without giving the bidders an appropriate basis to evaluate the risk.*
- *Balanced risk allocation between authority and private contractor; the private contractor should not be forced to assume risks which they cannot – by any means – influence (may sound trivial, but is not in reality).*
- *Reasonable reimbursements of Bid costs to the "losers".*

15.4.5 Australia

The conversation in Australia is more around how the industry will cope with the huge pipeline of infrastructure projects, many which would be classified as mega PPPs. A couple of the reforms being considered include:

- *Reducing the number of shortlisted bidders from three to two, thus reduce the number of resources (contractors, architects, engineering consultants) tied up in losing Bids.*
- *Better utilisation of augmentations for contractors that are performing well – i.e. if a Consortium has performed well in delivering a PPP, reward with exclusive negotiation of further works – example might be a schools PPP, or section of railway (Canberra Light Rail is a good example where the Gov't might end up having five or six stages, but have only awarded one thus far. Further sections could be awarded using augmentations rather than a separate PPP) – again, the challenge being that industry can't deliver this pipeline if everyone is tied up in Bids.*
- *In terms of Bid costs, we are seeing a greater level of Bid cost reimbursement by governments to encourage participation, but the challenge of resourcing is greater. We need to be more efficient with what we have, so reducing competition or reducing the level of effort required to get to preferred bidder is key.*
- *Change mechanisms are sometimes poorly conceived and drafted in project agreements so that even simple scope changes can become very complex and expensive during the operating period of the assets. More realistic contract terms are needed in this area.*
- *From a lawyer's perspective (and, I suspect, technical personnel too), the documentation has become too voluminous. Participants become too bogged down in detail and lose focus on issues that genuinely deliver value or address risk. I suggest that solutions to this include: (i) standardisation – the UK model has been very good at this, through a combination of development, a firm central hand to prevent individual projects departing from standard positions, and market acceptance; (ii) procuring authorities (and their consultants) should resist the temptation to generate ever lengthier documents; (iii) bidders and sub-bidders grow to accept the quirks in standard documents.*
- *Risk allocation is tough and can lead to bad outcomes, e.g. Carillion. Many risks are lumped on contractors that they can't really control. An easier risk allocation would be good but there are problems: (i) bank debt in PPPs means that consequences of delay are very significant. (ii) When the public sector gets smashed on claims, people come under a lot of criticism and scrutiny. So this motivates individual project teams to use the bargaining power of procurement to push tough risk allocation. The contractors don't help this; some of the claims are hugely inflated. I think we need: (i) recognition by the public sector that taking risk means sometimes risks eventuate and this leads to cost increases; (ii) politicians resisting the urge to use cost increases to criticise procuring*

authorities; (iii) contractors need to avoid killing the goose that laid the golden egg, i.e. be restrained about claims to lead to fair outcomes, not ambit claims that push the public sector into a corner.

Considering the above comments, we will now consider some of the main issues in more detail, on the basis of some working examples.

PPPs that are **politically driven for the wrong reasons** are often the ones that get into trouble and attract criticism. As stated before, there must be a definite need based on sound commercial principles. A significant case study in this regard is the Wanthaggi PPP Desalination Plant in Victoria, Australia. https://en.wikipedia.org/wiki/Victorian_Desalination_Plant

The construction cost AU$4.0 billion against an original government budget estimate of AU$2.9 billion, with the contractor losing over AU$500 million in the process. It was completed in December 2012, but not used for the first time until March 2017.

It is a 27 year PPP contract and the annual cost to the state is AU$608 million. Low usage is anticipated in the future.

The big questions are *'was it really needed?'*, and *'were there not much cheaper alternatives?'*

Certainly in 2007 when the project was announced, Melbourne, with its 4.7 million people, was facing critical water shortages, with the city's water storages at their equal lowest-ever level at 28.7% of capacity. When it was completed 5 years later the storages were back up to 81% capacity.

Was it politically inspired, without a convincing feasibility study? This question has been asked by many, given the above results and given that other schemes were simultaneously being implemented.

By comparison, the Kwinana Desalination Plant in Perth, Western Australia, was completed in 2006 with 30% of the output capacity of the Wonthaggi plant and at a cost of AU$387 million.

Another sector that has run in to some trouble in recent years, with a number of financial failures, is that of capacity revenue projects (toll roads, rail lines, etc.). However a significant change of approach has taken place – **many investors are no longer prepared to take the same levels of traffic volume and toll revenue risk** as previously. This is because of a number of financial failures due to lack of sufficient traffic, or because the tolls or rail fares have been set too high and have been rejected by users. In each case the tolls and fares were calculated on the basis of traffic projections by expert consultants and by the local traffic authorities, with these projections then being calculated into a toll or fare that would support the financial requirements.

In other words, *'financial engineering'* was used to make the financial model work without always taking into full account the public reaction to the level of the *'user pays'* tolls.

The alternative economic approach, which is now being increasingly widely used, is to procure toll roads not on traffic toll revenues, but on the *'availability'* model (similar to PPP social infrastructure – schools, hospitals, etc.), whereby the public sector pays the private sector provider a fixed annual amount that covers the cost and amortisation of debt and equity, and the costs of operations and maintenance, all out of user-funded taxes. There is a payment deduction regime if the unavailability of the motorway or rail line exceeds the agreed programmed maintenance closures.

In relation to toll roads, Australia has been a notable recent example in experiencing this problem, having got too bullish on toll roads from around 2000, with the following badly conceived and 'financially engineered' PPPs that became financial disasters.

- Airport Link, Brisbane – stakeholder losses + AU$3 billion (see Chapter 4)
- Cross City Tunnel, Sydney
- Lane Cove Tunnel, Sydney
- Clem Jones Tunnel & Motorway, Brisbane

Causes of failure on these projects included:

- Over-optimistic traffic projections – significantly lower traffic volumes than forecast
- Excessive tolls rejected by the public, which used alternative routes, meaning lower revenues
- Excessive total capital costs, with toll prices artificially set to service these costs.

In Germany the Herren Tunnel in Lübeck encountered a similar problem when motorists found that they could take an alternative route to the tolled tunnel at no cost. These experiences and other similar ones have been instrumental in a move towards using 'availability' models for new toll road projects around the world, although many projects and countries continue to use the traditional 'user-pays revenue' model for toll roads and rail.

A consistent complaint heard from private sector bidders across all countries concerns the very high level of Bid costs incurred in order to secure a PPP project. **Current Bid cost levels are generally considered to be quite out of proportion to the potential for return and the probability of winning a PPP project**. Furthermore the actual margins delivered on winning a PPP project are rarely sufficient for a concessionaire or contractor to be able to recover the lost Bid costs on their unsuccessful Bids. This is contrary to the commonly perceived public image that contractors and equity investors are 'milking the public purse' on these PPP engagements. Closer analysis of their total business model demonstrates how the financial rewards for the private sector are overall not really commensurate with the level of risk and high Bid costs that have to be assumed in securing and executing these long term, fixed price commitments. This lies at the core of many contractors' current reluctance to pursue PPP projects – there are better risk/reward profiles available to them in other markets.

So, there surely has to be a better way? One way perhaps would be for the public sector client to manage and firm up the initial design, and to then put the project out to competitive bidding. Currently, in most PPP Bid markets, we see the expensive and inefficient situation where, typically, three or more bidders will each spend a small fortune on preparing their own unique complete design, only to see all but one of these designs thrown into the rubbish bin at the end of the day when the winning Bid is selected. Governments argue that if they handled the design process then they would not get the benefit of innovation that comes out of a competitive situation. Many people disagree with this view.

> An extreme example of a huge waste of bidding costs was the Brisbane Airport Link project in Australia in 2009, on which three selected consortia reportedly each spent around $20 million on their respective Bids, with only one winner in the end. So $40 million of valuable company working capital (shareholders' funds) went down the drain. The largest percentage of this expenditure was for civil engineering design, yet realistically there were only a small number of different ways that the motorways, tunnels and bridges on the project could be designed.

Governments could readily engage a group of experienced design consultants to develop the outline concept for most PPP infrastructures – including hospitals, prisons, schools, toll roads, etc., and by doing this could still leave the door open for innovation in the bidding process, which could actually be a good incentive towards submitting a winning Bid. The cost savings would be quite significant.

By their nonrecourse, SPC nature, with numerous stakeholders and interested parties, **PPP projects are contractually quite complex compared to other forms of design and construction**. In the 1990s and early 2000s, this contractual complexity led to a lot of misunderstanding between the various stakeholders and it became a minefield for contractual disputes.

Today the suite of contract documents for PPP procurement in most established markets has become well refined and reasonably well standardised. Unfortunately it still happens that the principle stakeholders and their lawyers try to introduce unworkable 'bespoke' clauses into project agreements and subcontracts for new projects, which only delay procurement, increase overall cost and end up with further dispute. Some people just cannot 'leave well enough alone'!

The majority of participants in developed countries at least have a good understanding of how PPPs work, and experience is being rapidly gained by those nations now developing their own PPP markets.

Normally a project will have four primary contracts: the Project Agreement, Finance Agreement, Design and Construction Contract and FM Services Contract. Alongside these are numerous other contractual and supporting documents, such as the original Bid brief, scope of works, the design and equipment drawings and specifications, collateral warranties, etc.

With such complex arrangements there is a lot of personnel involvement from all stakeholders. **Therefore human behaviours are still one of the most important dynamics in this partnering between the public and private sector**. Whilst the parties may have many objectives in common, it should be recognised and respected that they do simultaneously have some conflicting interests. It is not just about contractual interface. Successful partnership working requires skilful and sensitive personnel handling by managers from both sides of the fence.

15.5 Efficient Structuring and Managing of PPPs

All areas of PPPs require effective human input in respect of:

– Collaboration
– Communications
– Relationship management
– Risk management.

These are the four key components of 'partnering' and they affect everything, including the concept and the community; approvals; the feasibility and contracts; design development and bidding; construction and commissioning; and O&M.

The different stages of a PPP project involve the following areas.

15.5.1   Structuring, Bidding and Finalising the Contracts and the Financing

- It is vital to understand the risk transfers and ensure that there are no gaps between the contracts.
- Services abatement risk transfers. Beware of caps in the FM services subcontract.
- Make sure realistic budgets cover three stages: D&C, commissioning completion and transition into steady-state operations. This is in the first two years of operations.
- The PPP partners, the lenders and the subcontractors must have a thorough knowledge of the contract documents, particularly on the obligations and risk transfers.

15.5.2   Establishing the Project Leadership and Team Spirit

- The SPC must provide strong project leadership that starts the project positively and proactively manages all key areas.
- Ensure effective communications and stakeholder relationship management and ensure relevant parties are never left out of the loop, but be careful not to be viewed as only a 'post box'.
- Negative communications can be very damaging – watch out for 'silent assassins'.
- The SPC should be the leader on reporting, expediting programmes, resolving issues, public relations, and stakeholder relationship building.

15.5.3   Agree the Key Processes with all Stakeholders at the Outset

The SPC needs to quickly put in place all the required processes and to get all stakeholders to buy in to them.

15.5.4   Construction Phase

- Design development lead time – don't start construction too soon on site just to 'make an impression'.
- PPP user groups – firmly reject unreasonable 'Xmas wish list' shopping by the government and/or user groups during detailed design development – to adjudicate this is an SPC leadership role. The issue is that at this stage the project agreement has been signed and the contractor has a finite contract sum to work within.
- Value engineering is the key to maintaining the contractor's margins, and the client's contract allowances, e.g. equipment procurement, and it is often necessary for the client to compromise and relinquish something in order to get something else. Effective value engineering mostly requires some 'horse trading' and 'give and take' to maintain the cost base and budgets, and it is very important that the FM services subcontractor is fully involved.
- Cost-neutral solutions keep everyone happy.
- Equipment selection and procurement can be a stressful area; keep everyone happy and maintain the budget.
- Make sure the FM services subcontractor endorses all drawings and equipment purchases so they cannot complain later.

- 'Fitness for intended purpose' (FFIP) or 'functionality' – contractual obligations in respect of FFIP can be a trap if these obligations are not clearly spelt out in the contract documents, because if they are not it leaves the door open for the government/operator to make unreasonable demands on the basis that *'it does not meet fitness for the intended purpose,'* which can become very subjective.
- Constantly monitor progress versus programme, costs-to-complete versus budget, and construction quality – be a step ahead of the independent technical advisors – you don't want the banks asking about something you should have reported.
- Don't let it become a *'remote'* site – the government contract is with the SPC and that is where they will call first if there are any issues – the SPC must be right on top of the construction and FM activities and pre-empt any short-cutting or lack of input; this also ensures the earliest knowledge of issues and hopefully makes it easier to resolve them.
- The builder must have an effective quality assurance (QA) system that ensures user group *'agreements'* are all picked up in the drawings and equipment schedules and incorporated in the construction, even before the amendments are re-issued – nothing annoys a client more than having to take the QA role by default through sloppiness in the design management.
- Major modifications/variations – stick to the process in the contract – generally the client has to pay for design and estimating new variations, and be careful of doing favours by incurring costs and providing *'free'* estimates for *'maybe variations'* that clients often request to save money; this practice has a habit of backfiring because the variations may not proceed.
- Subcontractors should also be careful to not put in rip-off prices because they think they have a captive client; it is a great way to damage the relationship and make it twice as hard to get genuine variations approved – the SPC leadership should jump right on this.
- Be very prompt with all contractual correspondence – efficiency builds upon itself.
- Modifications and variations are regular causes of trouble and damage to relationships.

15.5.5 Commissioning, Completion and Transition to Operations

- The single biggest issue in PPP projects around the world is that contractors do not allow enough time for commissioning; this is very different to normal D&C projects, which traditionally might only be 95% finished at practical completion and the remainder is completed during the defects liability period.
- The SPC needs to engage a highly experienced transition manager to liaise with and coordinate the activities of all parties – at an early stage – say one year from completion. This is not the same as a commissioning manager, who is supplied by the contractor.
- Ideally, all stakeholders contribute to cost of the transition manager, but agreement for this is difficult to achieve sometimes.
- Use a Deed of Variation or an Amending Instrument during design and construction to record all changes to the specifications in order to reduce the potential for disputes after commencement of operations. It makes the client focus on reviewing and checking detailed architectural and services drawings and equipment schedules.

Sometimes it can be difficult getting clients to agree to Amending Instruments as they feel they are giving away risk transfer – arguments in favour are:
- This is what the user groups want and the client has agreed – so lock it in
- Staff will change and you don't want expensive disputes with new staff that will damage the good relationship
- Risk still lies with the contractor for the facility to be FFIP, but is reduced
- The original brief will be obsolete – the Royal Women's Hospital in Melbourne had over 700 design and equipment changes – they need to be accurately recorded and signed-off by the stakeholders
- Better to do the hard work on agreeing the detail 'now'.
- In regard to the facilities management (FM services), remember that,
  - Thorough training = management competence
  - Management competence = no abatements (efficient preventative maintenance, use of cure plans, etc.)
  - Understanding this = reduced risk priced into Bids
  - Remember that with constantly changing staff, in regard to both the facilities manager and the client, it is important to build into the contracts on-going structured, regular training and accreditation courses
  - *Note that PPPs often provide opportunities during their life for the SPC to enhance the returns on investment, so always have a section for this in monthly reports.*

15.5.6 Defects Liability Period

- Management of defects. A major issue with PPP projects, especially hospitals and prisons, is that it is necessary to be 99.9% defect free at practical completion, for two reasons:
  - The independent tester will require it
  - Access for fixing defects will be restricted as soon as the facility goes into the operations phase.
- Use of *'task teams'* is a much more efficient and overall more economical process for everyone, because coordinating subcontractors is difficult; they are careless and damage the work of others without a care in the world. A typical task team consists of a supervisor from the contractor and one skilled tradesperson from each of the main subcontractors involved in the completion and commissioning, e.g. an electrician, plumber, painter, furniture joiner, IT technician, etc. The team works its way sequentially through rooms and corridors, fixing all defects (snags in the UK) and locking the rooms off as they proceed. Contractors should make *'task teams'* a mandatory procedure because it is a *'win–win'* for everyone – and it is really good risk management.
- Managing defects rectification becomes complicated and requires an interface agreement between the contractor and the FM services subcontractor, with the FM subcontractor being responsible for all but major items involving warranties, such as large equipment breakdown. If not handled this way there is unnecessary exposure to abatement risk.
- It is far less expensive and more efficient to rectify defects before project completion – just ensure there is enough time in the programme.

15.6   PPP Claims and Disputes

Even when the community need and/or commercial viability has been clearly demonstrated, PPPs (PFIs) still get into trouble for a variety of reasons, such as:

- Not undertaking thorough due diligence
- Not undertaking in-depth risk analyses
- Not engaging experienced and competent project teams (expertise is required by all parties)
- Not establishing efficient management platforms
- Not implementing effective collaboration, communications and relationship management.

15.6.1   PPP Disputes – Typical Categories

- Land procurement and boundaries – client responsibility
- Site access delays – client responsibility
- Design changes after financial close – client
- Construction quality deficiencies – SPC, transferred to builder
- Construction under-pricing – builder
- Equipment usage deficiencies – SPC, transferred to builder
- Renewal of rates and charges – client/SPC/FM provider
- Not fit-for-purpose – SPC, possibly builder
- FM services deficiencies – SPC, transferred to FM provider
- Construction delay damages – SPC, transferred to builder
- Services penalties – SPC, transferred to FM provider
- Maintenance, security breached, cleaning
- Staff levels, life cycle replacement
- Unavailability of facilities or roads

15.6.2   PPP Disputes

These occur because:

- The contract terms are ambiguous, or not specific enough
- The party claiming dispute did not properly understand the risk transfers and their obligations when they signed the contract
- The SPC has not met the programme, but is still liable to penalties, even if the problem is caused by subcontractors
- The builder has not met the specifications for construction quality or supply of equipment
- The FM services provider has not met the services delivery specifications
- The party claiming dispute has not made enough financial allowance and is trying to avoid the obligation
- The SPC, builder or FM services provider has been unable to meet the contractual programme
- The client is unable to deliver on their obligations, e.g. land availability when required
- Calculations for variations, services charges, renewal of contract rates and insurances, etc. are disputed.

PPP contracts include the transfer of many risks down the line and sometimes the terms for these are impractical and unworkable – and have onerous and unfair abatements (penalties) attached to them for non-delivery.

This can occur for different reasons, such as over-zealous government parties taking advantage of inexperienced private parties; inexperienced lawyers on both sides drafting 'bespoke' terms into the contracts, acting for similarly inexperienced government authorities, equity investors and contractors. Hence the importance of experience! If you aren't sure, seek advice.

Although PPPs have a suite of contracts, <u>the government client only has a</u> **<u>direct contract</u>** <u>with the SPC equity investor (concessionaire)</u>, and not with the construction contractor or the FM services provider, even though their subcontracts form part of the project agreement. Remembering this basic precept often helps to put claims and disputes into perspective.

15.6.3   Transfer of PFI Risk

The National Physical Laboratory project in the UK is well known because:
- The contractors were unable to deliver the required technical specification
- The losses on this project, along with costs overruns on the Millennium Stadium forced John Laing to sell its construction business for £1 to R O'Rourke & Son
  'Firms ran up £100 m loss on PFI laboratory scheme.'
Wednesday 10 May 2006 09.20 BST https://www.theguardian.com/business/2006/may/10/highereducation.businessofresearch
  See Chapter 4 for more details on this project

15.7   Summary of Key Factors for Success and Minimising Risk

- Understand the contracts – builders have a habit of not reading the project agreement and therefore do not fully understand the risks that are being transferred to them
- It is vital that all subcontractors understand the risk transfers
- Always involve the FM services subcontractor early on, from the start of the Bid
- Ensure that cost planning and design development work closely in parallel
- The SPC must provide strong leadership in all areas, promoting effective collaboration, communications and stakeholder relationships
- Establish effective management platforms for document control, activities schedules, etc.
- Ensure that all project documentation, correspondence, minutes, claims and variations, programmes, and schedules of key activities and key issues, etc. are kept up to date daily
- Manage issues proactively – *'talk first – write later'*; aim for the project to be dispute free
- Closely monitor progress versus programme and cost-to-complete versus budget
- Be on top of the risk management – *'anticipate, anticipate, anticipate'*.

Finally, one of the real benefits of PPPs is the partnering aspect and community participation. For a PPP to be successful, as the great majority are, then all stakeholders must genuinely embrace the community aspects of being respectful partners and adhering to open, honest, and constant communications. This often means a sea-change in attitude for some stakeholders, and for building companies in particular.

The common objective for all stakeholders, including the community, must be to deliver a first class facility in all respects, with everyone being delighted with the end result.

> **Successful PPPs reflect the personnel from all stakeholders collaborating, respecting and working smoothly together in all areas.**

Section D – Management of Risk

16

A Tale of Oil Rigs, Space Shots, and Dispute Boards: Human Factors in Risk Management

Dr Robert Gaitskell QC, Chartered Engineer, Arbitrator, Mediator, Adjudicator, Expert Determiner

Keating Chambers, London, UK

16.1 Human Factors in Risk Management

'As we boarded the helicopter in Aberdeen they asked for the name of our next of kin. We had already had a lecture on how to bale out if the helicopter went down en route to the oil rig.

As we approached the rig there was a strong breeze so that the helicopter floated about as it came in to land on the rig helipad. That was a pity, since this was a tension-leg platform rig, secured to the sea bed by strong cables, so it bobbed around like a cork. I prayed for the skill of the pilot as he attempted to put floating aircraft onto bobbing helipad. Then I noticed a sobering sight. On the edge of the pad was someone with the appearance of a silver spaceman, kitted in fireproof garments and aiming the long nozzle of a foam gun at us as we closed in on the rig. Plainly, it was his job to smother us with foam to quench any fire if we hit the rig – a fire on an oil rig is catastrophic.

I suddenly understood one key aspect of the human factors in risk management. Ensure good training of staff – in this case the pilot, and the aircraft mechanics that keep the helicopter in the sky. But also have a viable fall-back – in this case the foam gun.'

Risk management in the construction industry is the difference between success and failure; between profit and loss; between life and death. The industry itself is inherently risky. How then do we best address the human factors in managing risk? We have already looked at staff training. Next we will look at making decisions using the best data available. The final human factor we will address is minimising disputes on site by the use of a 'standing dispute Board'.

16.2 The Challenger Disaster

A second significant human factor in engineering risk management is ensuring that important decisions are made in the light of all the known information, and the decision is not skewed by an overbearing client or a bullying boss. A classic case study here is the 1986 Challenger disaster. Imagine yourself in the situation that a group of American engineers found themselves in.

Global Construction Success, First Edition. Charles O'Neil.
© 2019 John Wiley & Sons Ltd. Published 2019 by John Wiley & Sons Ltd.

It was a freezing cold night in Florida. At Cape Kennedy, the Challenger space shuttle was on the launch pad. Around 8.15 p.m., Eastern Standard Time (EST), 27 January 1986 a teleconference started. On one end of the line was NASA's rocket engineering establishment, the Marshall Center in Texas. On the other end was Morton Thiokol, in Utah. This company, Thiokol, had the contract to build the solid rocket boosters (SRBs) for the shuttle. For some time all concerned knew there had been problems with the joints between the segments of the boosters. In particular, in operation there were difficulties with the gaps that opened up in the joints containing the O-rings. A Thiokol engineer, Roger Boisjoly, felt there was a link between low temperature and damage to the seals from hot ignition gases. Various tests were carried out, but the results were inconclusive.

As the teleconference started, the NASA team were under pressure to launch. On board would be a school teacher, Christa McAuliffe, and it was intended that she would link up live with President Reagan as he gave his State of the Union address. Back in Utah the Thiokol team of engineers had already decided that the freezing conditions would reduce the O-ring resiliency unacceptably. They had agreed amongst themselves to recommend scrubbing the launch unless the O-ring temperature was at least 53 °F. The problem was that on site the temperature at launch was expected to be only 29 °F.

The telephone discussion did not go well. Thiokol gave its recommendation about 53°. Boisjoly referred to his concerns about low temperatures detrimentally affecting the O-rings. The customer asked him to quantify his concerns but the data was incomplete. Larry Mulloy of NASA was not impressed by Thiokol attempting to create a new criterion at the last moment. He and others thought the 53 °F limit was arbitrary. Larry burst out: 'My God, Thiokol, when do you want me to launch, next April?'

Eventually Thiokol requested a five-minute off-line private review. In a gloomy discussion lasting half an hour Boisjoly and other engineers explained their concerns.

The general feeling in the Thiokol team was that the engineering analysis was weak. The internal discussion was chaired by Senior Vice President Jerry Mason. He reminded everyone that the space vehicle was designed to be launched year-round, so that restricting it to launching only on warm days would be a serious change. He reiterated NASA's points. Although Boisjoly and his engineering colleague Arnie Thompson strongly fought their corner, the other engineers said nothing. Eventually, Mason said that if there were no more engineering arguments then Thiokol management would make a decision. Boisjoly and Thompson both stood up to make their points a final time, and Boisjoly showed the managers two photographs from earlier launches.

Present, besides the engineers, were four managers, all engineers by background. Their views were polled: it was three for a launch but Robert Lund was reluctant to support the majority. Nevertheless, after being told to 'take off his engineering hat and put on his management hat', he stopped resisting. The teleconference resumed and Thiokol approved the launch.[1]

As the world watched the television images and those at the launch site huddled in their coats on a bitterly cold morning, the space shuttle headed off at 11.38 a.m. Smoke soon started emanating through the O-rings. Soon there was a flame that hit the external fuel tank and the strut holding the booster rocket. The hydrogen in the tank ignited, the booster rocket broke loose, smashed into Challenger's wing then into the external tank.

---

1 Collins, H. and Pinch T. (1998) *The Golem at Large.* Cambridge University Press pp. 30–56.

At 50 000 ft Challenger was totally engulfed in fire. It had been aloft for only 76 s. The crew cabin separated and dropped into the ocean.

President Reagan set up a commission to investigate the accident ('the Rogers Commission'). This group included Richard Feynman, a theoretical physicist from the California Institute of Technology. In the course of a televised hearing he made sure that he had a glass of ice cold water in front of him. When the camera turned to him he produced an O-ring and dropped it into the freezing water. Then he pulled the ring out and snapped it, demonstrating that it had become less resilient at low temperatures. He later explained how, unlike most of the other Commission members, his investigation was not simply theoretical, and that he had spoken to engineers who had actually been involved in the project. He had strong views on the reliability of the shuttle and, in order to ensure these were included within the report, he threatened to withdraw his name from it unless those views were included within an appendix. He concluded: 'For a successful technology, reality must take precedence over public relations, for nature cannot be fooled'.

Roger Boisjoly left his job at Morton Thiokol and became a speaker on workplace ethics. He was particularly critical of intense customer intimidation. The American Association for the Advancement of Science awarded him its Prize for Scientific Freedom and Responsibility.

16.3 Dispute Boards

Dispute Boards (DBs) involve a procedure whereby a panel of three engineers/lawyers (sometimes just one) is appointed, often at the outset of the project. Ideally, the DB will visit site three or four times a year and deal with any incipient grievances. This often avoids a complaint crystallising into a dispute. Disputes festering on site are known to sap morale and generate an air of grievance over the whole project. The result is other disputes, strikes, and even accidents.

16.3.1 DB Background

After successful US experience with DBs in the 1960s and 1970s, in 1995 the World Bank made the procedure mandatory for all International Bank for Reconstruction and Development (IBRD) financed projects in excess of US$50 million. Since then the procedure has been introduced into many contract forms, particularly the FIDIC (the International Federation of Consulting Engineers) suite.

The commonly favoured model for DBs in the USA was and is the dispute review Board (DRB), under which 'recommendations' are issued in respect of the particular dispute being dealt with. This is a relatively consensual approach to dispute resolution. Broadly, if neither party formally expresses dissatisfaction with a recommendation within a stated period of time, the contract provides that the parties are obliged to comply with the recommendation. If either or both parties do express dissatisfaction within the limited time period, then the dispute may go to arbitration or court litigation. Although the parties may choose voluntarily to comply with a recommendation while awaiting the decision of the arbitrator or court, there is no compulsion to do so.

16.3.2 FIDIC DB Clauses

FIDIC, with World Bank encouragement, introduced the DB procedure into its engineering standard forms by way of the 1995 Orange Book form. This was followed by its 1996 introduction into Clause 67 of the Fourth Edition of the FIDIC Red Book for Building and Engineering Works Designed by the Employer. FIDIC adopted the dispute adjudication Board model, whereby effect must be given forthwith to a Board decision. If no *'notice of dissatisfaction'* is issued within 28 days of the Board's decision, it becomes final and binding. If a notice is issued then the matter may proceed to arbitration, although the parties are obliged to comply with the decision in the meantime.

A Board is able to fulfil two separate functions:

1) It can give informal assistance at an early stage with embryonic disagreements, simply by talking through complaints with the parties.
2) The Board may deal, on a more formal basis, with specific disputes referred to it, giving a determination as required.

16.3.3 Operation of a DB

Two types of DB need to be distinguished, although they may be used in conjunction with each other. There is the so-called 'standing' DB, where the Board is appointed at the outset of the project and is in place throughout, making periodic visits and dealing with complaints when they first arise, so that, generally speaking, they never develop into disputes. However, on some projects the parties only appoint an 'ad hoc' Board when a particular dispute arises. Essentially, this process is akin to simple statutory[2] 'adjudication' as widely used in the United Kingdom and various Commonwealth jurisdictions.

Where a standing DB is used there is an opportunity for the composite Board, consisting of members with both legal and engineering expertise, to walk around the site at regular intervals to see what work has been done. Thus, if subsequently a dispute arises about work that has been covered over, there is a chance the Board will have seen some aspect of that work before it disappeared. In addition to inspecting the site, a typical visit by a Board will involve holding a semi-formal meeting at which all interested parties may air any grievances that they have. The Board, of course, is entirely neutral and independent and will give all concerned a fair opportunity to explain their views.

The precise procedure for any particular DB will be set out in the contract governing the creation, constitution, and activities of the Board in question. It is sometimes the case that the contract will permit separate meetings to be held, so that the Board may meet with one or other party privately in 'caucus'. It is generally good practice, before this happens, for the matter to be discussed so that both parties know precisely what is contemplated, and can agree on the procedure to be followed. If the procedure is not consensual, then there is always the danger that one or other party will feel that things were said at the other party's separate meeting with the Board that ought not to have been said, or which are prejudicial to it and cannot be dealt with since the details are not known.

Sometimes the contract will provide that, after the site meeting and, ideally, prior to the Board departing, it should produce a short written report for distribution to all

---

2 In the UK the initial relevant statute is the 1996 Housing Grants etc. Act. Now see the Local Democracy, Economic Development and Construction Act 2009.

concerned. This would record the attendees and details of the meeting, what was seen on the site visit, any grievances raised, and any determinations (whether recommendations or decisions) made. Future action required of the parties (e.g. the production of documents for the next meeting), and an explanation of what the Board itself will be doing, if anything, prior to the next meeting, should also be noted in the report. The proposed date for the following visit can also be identified.

Ordinarily, a standing Board will remain in place until the conclusion of the project, which is often marked by, for example, a certificate of practical completion or some equivalent document.

My experience of standing DBs is that they work; they really do stop all or most complaints crystallising into disputes that need expensive arbitration or litigation.

An exciting project that is using a standing DB is the International Thermonuclear Experimental Reactor (ITER) Fusion for Energy project in the south of France. This aims at demonstrating electricity generation from 'fusion' rather than from the conventional fission process. This high-value experimental project is supported by a wide range of nations. I ought to emphasise that what I say is all in the public domain – on the project's website, or elsewhere – so you will find no commercial secrets in what follows!

16.4   Nuclear Fusion

Nuclear fusion is quite unlike nuclear fission. The fission process has, of course, far more public recognition, since that technology lies behind the atomic bomb and conventional nuclear power stations. In the fission process the nucleus of an atom splits into smaller parts, usually producing free neutrons and releasing a very large amount of energy. In a conventional nuclear power station a nuclear reaction is deliberately produced. The fuel rods are bombarded with neutrons and the result is that further neutrons are emitted. This sets up a self-sustaining chain reaction that releases energy at a controlled rate in a nuclear reactor (for a power plant), or at a very high uncontrolled rate (in an atomic bomb). The fission process is linked with nuclear waste problems and with well-known examples of escaping radioactivity (e.g. Chernobyl).

The fusion process, by contrast, is significantly different. At the moment we know a limited amount about the process and are on a steep learning curve. Much of what we do know is very encouraging. Stars, including the sun, experience the fusion process at their cores. The ITER project aims to establish, if it can, that the fusion process may be used to generate electricity on a commercial basis. The process that is to be used involves two fuels that are relatively easily obtained. Deuterium may be extracted from seawater, and lithium is in the Earth's crust. Used together in the fusion process they create tritium on a significant scale. Ultimately, therefore, the supply of tritium is potentially unlimited. Mass for mass, the tritium/deuterium fusion process envisaged for the ITER Project is expected to release about three times as much energy as uranium 235 fission. Of course, this will be millions of times more energy than any chemical reaction such as burning fossil fuels like oil, gas, or coal.

The ecological credentials of fusion include the fact that it emits no pollution or greenhouse gases. Its primary by-product is helium, an inert, nontoxic gas. Unlike fission *'melt-down'* chain reactions, there is no possibility of a fusion *'run-away'* reaction, since any alteration in the conditions of a fusion reaction results in the plasma cooling within seconds so that the reaction ceases. There is a low waste output.

16.5 The ITER Project

ITER originally stood for the 'International Thermonuclear Experimental Reactor'. However, nowadays it is generally taken to refer to the Latin word for 'the way'. The member states for the project are the EU, India, Russia, China, South Korea, the United States, and Japan.

Broadly, the ITER Project involves about 10 years for the construction of all facilities at Cadarache in southern France, followed by 20 years of operation. If this essentially experimental project is successful, then a demonstration fusion power plant, named DEMO, will follow, introducing fusion energy to the commercial market by converting the heat generated by the fusion process into electricity in fairly conventional ways familiar to those with an understanding of current power plants.

The broad objective of the ITER Fusion Project is to establish that the reactor, using 50 MW of input power, is able to produce 10 times as much (500 MW) of energy output. Provided that can be achieved for a relatively short period (a matter of minutes) then the principle will have been established and the ultimate success of the DEMO power plant is assured. Another of the key objectives of the ITER Project is to verify that tritium, one of the necessary ingredients for the process, can be 'bred' in the reactor, so that the supply of that fuel becomes self-supporting.

The technological and scientific challenges involved in the ITER Project should not be under-estimated. Pierre-Gilles de Gennes, the French Nobel physics laureate, once said about fusion: '*We say that we will put the sun into a box. The idea is pretty. The problem is, we don't know how to make the box*'.[3]

16.5.1 Plasma

To be successful, the reactor must contain high temperature particles, with their enormous kinetic energy, in a sufficiently small volume and for a sufficiently long time for fusion to take place, creating the plasma. Ordinarily protons in each nucleus of the isotope fuel will strongly repel each other, since they each have the same positive charge. However, when the nuclei are brought sufficiently close, with sufficient energy, they are able to fuse. In the ITER tokamak machine the nuclei are brought close together using high temperatures and magnetic fields.

The plasma in the ITER tokamak is a hot, electrically charged gas. It is created, at extreme temperatures by electrons separating from nuclei. About 80% of the energy produced in the plasma is carried away from the plasma by the neutrons, which, having no electrical charge, are unaffected by the constraining magnetic fields. These neutrons then hit the surrounding walls of the tokamak, and are absorbed by the blankets on the walls and so transfer their energy to the walls as heat. In the ITER Project this heat is simply dispersed through cooling towers. However, in the forthcoming DEMO fusion plant prototype the heat generated will be used to produce steam and, through the intermediaries of turbines and alternators, generate electricity.

---

3 Wikipedia: 'Reactor overview'.

16.5.2 Magnetic Fields

The plasma needs to be heated to 150 million degrees centigrade in the core of the machine. Plasma at that temperature, and with its constitution, cannot be allowed to touch the walls of the reactor, since the plasma would rapidly destroy any constraining vessel and would also cool down, ending the process. Therefore, the plasma is controlled by so-called *'magnetic confinement'*. The plasma is shaped by magnetic fields into a ring, or 'torus', and thus it is kept away from the relatively cold vessel walls. These surrounding walls have *'blanket modules'* containing lithium. They are termed *'breeding blankets'* because, as part of the fusion reaction, tritium can be generated.

16.5.3 The Tokamak Complex

At the heart of the project is the ITER tokamak machine. This machine draws on the experimental work that broadly stems from a major breakthrough in 1968 when Soviet Union scientists achieved temperature levels and plasma confinement times significantly beyond anything achieved hitherto. The Soviet scientists termed their device, which achieved doughnut-shaped magnetic confinement, a 'tokamak'. The ITER tokamak will be twice the size of the largest current machine. If all goes according to plan, fusion power should be feeding into the world's electricity grid systems by about 2040.

The ITER construction contracts may well make good use of the DB procedure. Certainly, this procedure has the potential for minimising disputes and dealing with crystallised disputes in a cost-effective way. Risk management involving this sophisticated dispute avoidance procedure is likely to pay for itself and bring a range of benefits in levels of satisfaction among the key players. This will contribute to the timely and efficient completion of one of the world's most exciting energy projects, and benefit us all.

16.6 Conclusion

We have looked at the human factors in risk management from three different perspectives: staff training, decision making using the best data available unimpeded by difficult clients or management, and avoiding and dealing with disputes efficiently. Of course, there are a multitude of human factors relevant to a proper risk management methodology, but the essential thing to bear in mind is that the more staff and workers are empowered by training, protection from bullying, and protection from intercompany disputes on site, the more successful the project will be.

17

Effective Risk Management Processes

17.1 Introduction

We know that construction and civil engineering comprise one of the world's biggest industries; a major employer, the generator of vast revenues and an economic barometer. Internationally construction averages around 5% of GDP, with countries ranging from around 2.5% up to 10%, so it is a very significant part of every national economy.

However, while the industry has made giant strides technically over the last few decades, in equipment, materials, and techniques, *it has surprisingly not made sufficient advances in anticipating and managing risk, both project and corporate.*

All over the world too many projects are exceeding their budgets and their planned timetables, and do not meet their facility objectives, despite all the hard-earned lessons of the past. This applies in developed countries as well as emerging ones; it happens with major international companies, medium sized domestic companies and government authorities.

Risks and Opportunities
The role of a risk manager should not be confined to anticipating and assessing the potential down-side risks of a project, but they should also be constantly looking for opportunities within the scope of the contract to enhance margins. In their initial reports during bidding and in their monthly reports after winning the project they should always have a section in which they report on opportunities. With private public enterprises PPPs especially, it is common during the operational life of a project for opportunities to arise that will enhance the revenues and margins.

17.2 Effects of Human Behaviour in Risk Management

A significant reason for this failure to properly manage risk is an inability to effectively manage human behaviour and the influence it can have on corporate and project risk management.

Human behaviour certainly has an impact on virtually every aspect of the construction industry and this of course is most often an underlying issue when risk management has failed, coupled with a breakdown in communication somewhere.

Global Construction Success, First Edition. Charles O'Neil.
© 2019 John Wiley & Sons Ltd. Published 2019 by John Wiley & Sons Ltd.

Risk management processes are meant to anticipate, intercept, and eliminate or mitigate detrimental human behaviour as part of their role, but clearly this does not always work.

When I published the forerunner to this book in 2014, *Human Dynamics in Construction Risk Management*, I stated then that there was surprisingly little material available on the effects of human behaviour in construction, just a few interesting research papers, occasional short articles in some industry newsletters, and no books that I could find.

Since then, however, there has been a surge of interest in improving risk management and the associated human dynamics, which is very encouraging but still nowhere near robust enough, given the number of construction companies still getting into financial trouble and causing serious knock-on effects to the supply chain.

So why does erroneous *human behaviour* interfere so often with management processes to the detriment of a project?

Simple answer – people are people! This happens at all levels, from the assistant surveyor or site manager, to estimators, commercial and legal managers and advisers, with senior management and all the way through to the boardroom. But there is no doubt that the really major problems emanate from the top and throughout this book we provide analyses to demonstrate the issues and recommendations to try and address the problems.

We place considerable importance on the interaction between the human inputs into projects and the communication processes because **when there is erroneous human behaviour it is often accompanied by a breakdown in communication, which results in this detrimental behaviour not being identified either early enough or not at all.**

17.3   Typical Project Risks

Risk management covers any potential issue that is likely to jeopardise the project at any stage of its life cycle. Essentially those issues that are typically the most vulnerable to *'human'* related faults are the following.

17.3.1   Client, Contractor, and Other Stakeholders

1) Creation by the client of an unrealistic output specification and financial feasibility.
2) Failure by the client to put proper financing in place.
3) Failure of the project sponsors to attain full 'buy-in' collaboration from all the stakeholders from the outset.
4) A failure by top management to have good people management and personal motivation, team spirit development, and ensure employee satisfaction.
5) Corporate social responsibility; community and political considerations – ignore these at your peril.

17.3.2   Project Bidding

6) Overly optimistic contractor bidding, with under-pricing combined with a failure to understand the full output requirements.
7) A failure to manage the Bid diligently, or just plain incompetence.

8) Authority approvals – a failure to adequately identify all that are required and have them approved in a timely manner can cause huge delay, leading to loss of income, liquidated damages, loss of confidence and credibility, decaying staff morale, etc.
9) Failing to engage and seek the advice of the future facility management FM services contractor at an early stage.

17.3.3 Award of the Contract, Design, and Construction

10) Having won the project, a failure to create a cohesive project team and integrate it. One of the worst problems that we give ourselves is to set out on a new enterprise (huge in its size and complexity) with new groups of people and organisations, with no pre-agreed plan for the management systems to be put in place and run; it's a recipe for disaster. Having no continuity of personnel from the Bid team to the project team, and therefore no continuing responsibility, can also cause all sorts of issues.
11) Health, safety and environment (HS&E) – all have greatly increased importance globally and failing to recognise this can be very expensive, both in physical and monetary terms.
12) Planning and programming, including failure to develop the programme in detail from an early stage (e.g. plugging in durations that turn out to be unrealistic because the component is far more complex and critical than envisaged).
13) Resource allocation generally – suitable key personnel, consultants, plant and equipment, e.g. a lack of competent people, or staff shortages, or staff who do not have the skills to recognise issues arising, or are not prepared to take responsibility.
14) Failure to immediately implement effective communications protocols, which are essential for success, otherwise the stakeholder and business relationships will suffer, together with reputations.
15) Not running design development and cost planning in close parallel after winning a project. This is a constant and major cause of cost over-runs.
16) A failure to make best use of the available 'new' technology programmes that increase efficiency, such as BIM (building information modelling) and IDF (intelligent document format).
17) Contractor failure to produce realistic budgets and control of costs and margins, maybe to conceal the likely truth in order to get approval, or maybe through inefficiency or incompetence.
18) Poor cash flow forecasting and management and unrepresentative financial reporting.
19) Failing to get consultants, subcontractors, and suppliers to provide their delivery programmes and their aligned progress payment requirements.
20) Unrealistic and unrepresentative project reporting, with project managers failing to speak up, or unwilling or unable to recognise and reveal problems, or **the misguided intent of keeping quiet and fixing problems at project level alone** (this is a particular problem that occurs regularly and risk management reporting needs to be designed to eliminate it). *Many project managers fail to recognise that by reporting a potential risk they obtain the assistance of the wider management team and no longer carry the worry alone,* and possibly jeopardise their job by not reporting it. Mostly it is an ego driven problem.

21) Failure to manage the work properly, which can lead to failure to adhere to required standards and practices, e.g. HS&E, quality assurance, use of codes of practice, etc. A lack of supervision and management can send the wrong message to staff and fail to identify problems and catch them before they become a disaster.

22) Project meetings – a lack of scheduled mandatory meetings with specific purposes, involving all stakeholders and with accurate minutes distributed promptly. The key emphasis here is the challenge of getting all the right people in the right place at the right time. This is often difficult when dealing with multiple parties, but failure to have a fixed meeting schedule results in the absence of key people and communication gaps. BIM provides a significant means of assisting with this.

23) Quality assurance (QA) – failure to diligently and methodically conduct continuous QA inspections and reviews, and a failure to understand the significance of 'doing it right the first time'.

24) Issues and claims – failure to proactively manage them can turn them into disputes, quite unnecessarily and at considerable time and cost.

25) Corporate governance breakdowns, e.g. breaches of bank covenants.

26) Task management – one of the keys to eliminating risk is to have an automated system for generating/reminding people of tasks that are due or close to due, e.g. Affinitext Task Manager – ensuring contract obligations are met on time and not overlooked.

In all the above risk management areas, clients and contractors mostly put the processes in place, but it is the failure of people to take responsibility and to do what is needed that allows disasters to develop. Given the above list it is beyond question that managing human behavioural issues is a vital part of risk management.

17.4   Keeping Risk Management Simple

One of the main reasons that motivated me to put together my previous book and this one is that I just do not understand why many people in the industry find risk management so difficult and do not do it properly. Perhaps they have not suffered the consequences, such as losing their own money or their job as a result of poor risk management leading to a bad project, or they are just not passionate enough about helping their employer to avoid risk and deliver a great project.

Or is there too much employment specialisation, with people not wanting to take any interest or responsibility outside their silo (refer to Chapter 11), or not being allowed to? Maybe they are in the unfortunate position of working in an organisation that does not have an open culture that allows them to speak out.

Most likely though the main problem is a failure in training, in not having a thorough understanding of why diligent, anticipatory risk management is so important. This is the fault of senior management.

I have been delivering projects successfully since the 1970s and for the first half of my career we didn't even use the term *'risk management'*, but our project management methods accommodated it nevertheless. As a specialist process, risk management did not really come into vogue until the 1990s. Nevertheless, we did carefully consider all the possible risks when designing and constructing projects and accommodated them in either the pricing or in the terms of contract.

So what is risk management all about, in practical terms? It is about sitting down at the start of a project, right from the decision to Bid, and visualising and thinking through the entire project, the political climate, the planning and approvals, the design and construction process, and asking yourself what activities might go wrong that need to be accommodated in the pricing, or dealt with by another stakeholder better suited to handle it, e.g. the client or a specialist subcontractor of some sort, or by taking preventative measures in advance to stop it happening.

We used to spend hours on our *'anticipation'* lists, thinking about all the things that could possibly go wrong and then putting in place actions to prevent or mitigate the particular issues if they should occur, or negotiate with the client to take on those risks if appropriate.

It is not difficult, but you have to think through every step of the project from the concept and planning; the involvement of all parties; the design and shop drawings; the site establishment and the construction processes; the health, safety, and environmental requirements, right through to the completion and commissioning.

Where there are areas of concern you must discuss it with the relevant stakeholder. It is too late three years down the track, such as happens sometimes with PPPs where the services subcontractor is not properly engaged in the bidding process, even though they will be responsible for the ongoing project for years to come.

Easy you say, but only for small projects! Wrong! It is no more difficult for a $1.0 billion project, except that with large projects it necessarily becomes a team effort, which requires structure and discipline (refer to the team structure in Figure 17.1).

The principles remain the same – visualise every step and *anticipate, anticipate, anticipate*!

The Best Form of Risk Management Is Constant Anticipation and Conscientious Efficiency in All Areas by All Personnel

With our *'performance contracting'* approach my company undertook numerous projects that were very successful for both our clients and for us. One really interesting case study was a series of contracts that our steel fabrication division had for the supply, fabrication, and site erection of the structural framework for 12 regional shopping centres in New South Wales and Queensland, Australia, with standardised designs and a building size for each of around 20 000 square metres. Between the first and the last of these, over a period of four years, we made significant gains in efficiency that gave up to 25% savings in time and enabled us to reduce the costs for our client, but also increased our profit margin substantially.

A key factor in this success was our 'anticipatory' risk management combined with real efficiency in all areas, which is actually the best form of risk management.

We had a marvellous team and team spirit on these projects and it is great when your work is both satisfying and rewarding for everyone.

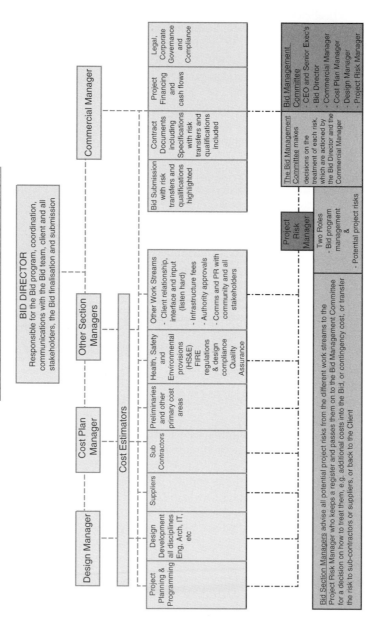

Figure 17.1 Bid Risk management team structure.

17.5   Procedures to Eliminate, Mitigate, and Control Risks

We recommend that the following items be closely considered.

An *'anticipatory'* early warning process should be used to alert the senior management team at the first indication of a possible risk, so that preventative or mitigating measures can be put in place. This process should be applied at all stages of a project – from the initial concept, through the tender stage, design and construct (D&C), and during on-going operations in the case of PPPs.

From the outset, with projects of all sizes, it is important to identify and schedule all potential risks. This should be a key element in undertaking project feasibilities and planning and in preparing Bids. The various risks can be weighted for probability and harmful value, and if considered necessary contingency allowances can be included in the cost build-ups. Remember, of course, that all potential risks are unlikely to happen on one project, so making a cost contingency for all of them is likely to make the feasibility unrealistic and not viable or make the Bid uncompetitive.

There are plenty of typical risk schedules and software programs available that include probability and cost contingency weighting. We have not included samples because this book is not meant to be an operations manual; it is sufficient that we cover the principles and importance of having a robust risk management process.

However, we have included a team structure (Figure 17.1) that shows how to structure the process for a major project, with section leaders reporting to a dedicated risk manager.

Risk management and key issues reports should be produced monthly or more frequently if required, for all ongoing projects. Risk management reports should be designed as checklists to assist managers, but experience in recognising situations at an early stage and common sense in addressing them are fundamental. Managers should be required to report **all potential risks** to the investors.

The key requirement is to identify potential risks at the earliest possible time so that they can be either dealt with promptly and eliminated as a risk or mitigated to the maximum extent possible. This requires an acceptance of the principle by site managers and those to whom they report; this can be more difficult to achieve than you would imagine.

The reason this is difficult is, of course, human nature. Often managers will take the view that they will solve the problem quickly so they don't have to alert the potential risk to those higher up. They seem to think, quite wrongly, that the early identification and reporting of a potential risk is a reflection on their capability personally and their sense of propriety is offended, so they keep the problem to themselves and try to resolve it before the next monthly risk management report – a big mistake! They overlook or do not understand that once they report the issue they are making life easier for themselves by sharing it with others and that they then have the combined weight and experience of the company to help them sort it out. It is very common problem, a mind set problem.

17.5.1 Project Executive Summaries for *'By Exception'* Reporting

So, the question is: 'on a major project, how can senior management be sure that the site team is being really diligent and efficient in their overall management; that they are reporting things factually and accurately; and that they will comply with the principle of "early warning" risk management?'

There are simple and effective processes that can be adopted that should enable senior management to be able to sleep at night.

All construction companies have standardised reporting and control formats for their projects – financial, operational, programme, quality assurance, health and safety, corporate governance, environmental compliance, etc.

By combining these formats with specific project parameters, the next step is to implement *'by exception'* reporting on a disciplined monthly basis, preferably signed off jointly by both the project manager and his or her supervising manager. This is effective *'early warning'* reporting as well.

There are simple proprietary formats available for this, or it is easy to develop your own (see Table 17.1), but surprisingly many construction companies do not use them. These formats are based on drawing up a list of key areas to be monitored, and if there is a risk issue or potential issue with any of these areas then this should be immediately notified by the project manager to the next level and also entered into the monthly report, with

Table 17.1 Example of a 'by exception' report.

Confirm that	
The direct construction costs to date are within budget	Y
The direct construction costs to complete are within budget	N details attached
The preliminaries and overheads are within budget	Y
The design consultants' costs are within budget	Y
Progress to date is in line with the programme	Y
Progress claims and payments are in line with the programme	Y
The quality assurance has received all third party sign-offs to date	
Lender's technical advisor	Y
Client's technical adviser	Y
Design consultants	Y
All required statutory approvals have been received	N details attached
All insurance policies are in place and premiums paid	Y
All health and safety requirements are being met	Y
All environmental compliance matters are in order	N details attached
All corporate governance items and statutory accounts are up to date (refer to the check-list)	N details attached
All contract documentation processing is up to date and not outstanding for more than 10 working days	Y
Confirm that there are no detrimental contractor or subcontractor human resource issues outstanding at the moment	Y
Confirm that there are no detrimental communications or relationship issues between any of the stakeholders	Y

details. It certainly should not be left until the next monthly report for notification. There are two strong reasons for immediate notification:

- It allows the maximum amount of time to prevent or mitigate the risk as much as possible.
- The project manager will have team support to handle the issue *('two heads are better than one')*. It becomes the company's problem, not just the problem of the project manager.

The monthly *'by exception'* reports can include the typical items shown in Table 17.1 for monitoring, with boxes for ticking in the affirmative, or green and orange traffic lights. The operations manager supervising the project should jointly sign-off the monthly report with the project manager.

This executive level *'early warning'* risk management format is designed to be easy and fast to complete and thus suitable for multiple projects being run by large companies. Strong discipline is required in the preparation of the aspects such as budgets, targets, programmes, and protocols; honest and accurate completion by project managers and their supervisors is essential.

> Really Effective Risk Management
> is all about
> Having the Right People in the Right Positions
> Anticipating, Anticipating, Anticipating
> and being
> Ahead of the Game.

17.5.2 Audit the Process Regularly

Most companies do have an internal audit process and these audits should be done on a regular basis, say quarterly, to ensure that the monthly reporting is being accurately and honestly carried out; however, internal auditing does have its drawbacks and these are examined in Chapter 27.

A regular independent audit is also a cost-effective way of providing backup to the internal reporting and auditing process. Unfortunately it is sometimes hard to convince the senior management of many construction companies that this is a prudent method of risk management, that is until they suffer a big loss on a project that could have been prevented had it been picked up earlier.

Corporate compliance and obligations diary. With every project there are a significant number of reporting and compliance obligations that should be undertaken on a regular basis and it is useful to have an annual compliance and obligations diary to ensure that the necessary work is prepared in advance and deadlines are not missed. The diary can include all matters and reports pertaining to the following corporate responsibilities:

- External stakeholders
- Statutory returns
- Annual accounts, audits, and tax returns
- Insurance premium renewals
- Securities institutes

- Licensing
- Local government
- Health and safety
- Environmental
- Training and apprenticeship schemes

The Vital Importance of Complying with Health and Safety Regulations

Anyone who has ever been involved with a fatal accident on a construction site will never forget the trauma suffered by the family of the person who died; the sense of loss and de-spiriting effect on the fellow workmates; and the stress of going through lengthy coroner's enquiries and in many cases court actions as well.

It is not difficult to establish the necessary routines and comply strictly with the minimum standards that are in each country. In fact many multi-national companies have minimum standards of their own that are superior to a particular national standard.

Ignore HS&E or administer it sloppily at your peril. It can ultimately put you out of business. Ensure that your procedures allow for regular independent checking on compliance.

Insurances Register. A register should be maintained that includes the policy details and premium renewal dates of all insurances relating to the group and the relevant projects, clients, third parties (authorities, subcontractors, etc.).

17.5.3 Collaboration, Communications, and Relationship Management

We recognise that renewed emphasis is being placed by many contractors on improved collaboration and communications in order to minimise risk and potential disputes, with an important part of this aimed at negating detrimental behaviour and decision making. There is now widespread acceptance of this need and a number of programmes are in place or being developed in this regard by both companies and industry organisations.

'*Collaboration*' is a vital function, but it has been well covered in other chapters so we won't elaborate on it further here, other than to reiterate that BIM is a great step forward in improving both collaboration and communications in addition to its technical benefits.

'*Communication*' is a vital word – it is a fact of life that whenever there is a problem it can invariably be traced back to a breakdown in communication, or reporting somewhere that overrides the risk management controls. This is a human problem and the best systems and processes in the world will not overcome it. The challenges of communication are demonstrated even in the writing of this book, with standardisation of terms, e.g. with some authors using '*schedule*' and some using '*programme*'.

Communication and relationship management comes in many forms within a construction company and between all stakeholders in projects, including:

- Meetings, reports, and accounts, correspondence and emails, documentation, training, public relations, and community liaison.

When a strong-minded project manager overrides or ignores fundamental issues such as compliance with fire regulations, it still amounts to a breakdown in communications because either the reporting systems or other stakeholders have not alerted responsible personnel further up the chain of command. As another example, mid-level managers can sometimes be a problem if they have a misguided ego-driven sense of power and enough autonomy to block communications up and down the company's communication and reporting chain. It can sometimes be difficult to convince headstrong, stubborn project managers that they should be identifying, capturing, tracking, and mitigating risk.

Protocols for effective communications should be established at both company level and project level.

17.5.3.1 Company Level
- Clear communication processes should be established by head office and regional directors in order to ensure that the company is not put at risk because of ineffective communications or communications that go to the wrong people.
- Email protocols are particularly important – misdirected emails can be highly embarrassing.

17.5.3.2 Project Level
A project communications protocol (PCP) should be prepared by the project director as an initial priority, agreed with all stakeholders, and written up and distributed.

The PCP must include:

- Clear communications protocols between client and contractor and with suppliers, subcontractors, and all stakeholders.
- A community relationship liaison and management plan.
- A crisis communications plan that includes all stakeholders and the community.
- An internal rapid escalation plan in the event of a crisis.
- Codes of conduct.
- Details of the communications platform(s) to be used, e.g. SharePoint, Team Binder, etc.
- Details of relevant compliance/corporate governance communication requirements, particularly in regard to safety.

17.6 Conclusions

Stringent risk management and corporate governance processes should be fundamental to any corporate management process and therefore be given top priority by senior management.

All the above processes are implemented by people, so that is your number one risk. This is why many successful construction companies insist on key decisions being signed off by two senior people.

It is clear from reports from all sectors of the industry that the value of efficient and effective risk management is not well understood by many contractors and its benefits are badly underestimated, despite all the publicity that it receives.

It is being poorly implemented by many clients, contractors, and consultants, and they are obviously not making best use of the available processes.

18

Risk Management and its Relation to Success in the North American Context

John McArthur

Chairman, Kiewit Development Company, Toronto, ON, Canada
Kiewit Development Company, Omaha, NE, USA

18.1 Introduction

My recent background has focused on Public Private Partnerships so my comments and views come from this perspective. Additionally, the comments and approaches that follow here are based on the premise that securing the project is a result of a highly competitive process. Personally, I have held senior positions in engineering firms and construction companies. For the past 14 years, I have established North American businesses for two different infrastructure investor/constructors. During this period, I have bid over 40 P3 projects and had success in 12, totalling a capital value of US$10 billion. Both wins and losses lead to a vast library of lessons learned, giving invaluable insights into the process of risk management.

Every project progresses through several stages and typically they are:

1) Initial strategic approach to bidding.
2) Project identification – is there an opportunity that suits capability.
3) Teaming – what composition of team members will meet the requirements of the qualifications stage in order to get shortlisted.
4) Detailed development of the Bid leading to submission (design and construction approach as well as the operations and maintenance plan).
5) Implementation.
6) Operations and maintenance

Risk management practices should be developed and implemented for each of the above six stages. A fundamental to be followed is the involvement by senior management at the starting point of each stage and their continuing supervision and involvement during each stage. In our busy world, too often senior managers delegate too much authority down. This results in the absence of intuition and experience that seasoned professionals bring to transactions.

Project stages – several key risk management points or questions:

1) Strategic approach to bidding:
   a) Does the opportunity fit with the company business plan?
   b) Does the company have the resources to deliver on the ground?
   c) Will the opportunity compromise the financial position regarding bonding, Letters of Credit, etc.?
2) Project identification:
   a) Ensure a successful track record of past implementation.
   b) Institute a project review committee independent of the project manager for project selection.
   c) Are joint venture partners available if required?
3) Teaming
   a) Do we need partners?
      i) For resume
      ii) To beef up staffing
      iii) To spread execution risk
      iv) To spread pursuit costs
      v) To bolster bonding capacity.
   b) What is the track record of potential partners?
      i) Dealings with our company.
      ii) Dealing with agencies.
      iii) Dealing with advisors, lenders, and legal firms.
   c) Are potential partners culturally aligned – corporately?
      i) If we have worked together in the past did we have a similar investment and construction approach.
      ii) Were we able to compromise on issues?
      iii) Did they contribute to innovation in investment and construction approaches?
   d) Is the split of responsibilities aligned with the share of the joint venture?
      i) Will the partner contribute enough experience resources to pull their weight?
      ii) Are we aligned as it relates to fees, margins, etc?
4) Detailed Bid development
   a) Two fully independent internal estimates are essential.
   b) Is the Bid development budget sufficient?
   c) Is senior management involved in strategy development from the outset and frequently throughout the process to Bid?
   d) Is there a clear Bid management and communication plan agreed and adhered to by all parties?
   e) Is there a schedule of regular team meetings where all parties will be given direction and have responsibility for reporting progress?

Risk management is a process, not a single activity or event. Avoiding the identification of risks, resulting in lack of their effective management, is the most common cause of projects not being fully successful, or failing completely. The process involves looking back, analysing historical information, drawing conclusions from this process, and then putting into place measures embodied in the process of risk management. Conclusions drawn from an open and objective review of past projects, both the successful and unsuccessful ones, will provide the ingredients for a risk management process. It is often best to involve senior management that were not involved in the past projects that are being reviewed. It is difficult to provide objective critical analysis of projects that one has been directly involved in managing. Team members that have been part of an unsuccessful project in the past have difficulty seeing what went wrong – they can't see the forest for the trees.

Successful risk management comes down to the human factor. Regardless of how detailed the plans for managing risk, or how integrated into the design and construction process they are, effectively managing the people that are required to deploy these plans will determine ultimate success. This means understanding the various personality types involved in a project and using the best management methods to bring out the optimum performance in people. This must of course be done in the context of a comprehensive project management plan (PMP). In an ideal setting, where project specific teams are drawn from extensive human resources, combining personality types representing a range of compatible strengths will result in the most productive and cooperative teams. When the pool of potential resources isn't large, it could be effective to have team members formally determine (e.g. the Clifton Strengths) their various strengths (and less strong strengths – formerly known as weaknesses) so that team members have some insight as to why people behave the way they do. Understanding why people behave in certain ways can make team member interaction more effective.

Risk management in North American projects is generally similar to other parts of the world. Probably the key difference in the United States results from there being a history of litigation arising from construction issues. This is less the case in Canada. There is a general propensity to take legal action to compensate wronged (or perceived to be wronged) parties. For this reason many larger construction companies have significant internal legal departments and equally sizeable in-house insurance and risk management teams. With possible litigation looming, in-house insurance and risk management teams have developed sophisticated risk analysis processes. As a result, they have considerable involvement in initial and ongoing project decision making.

It is common for the parties to end up in lengthy court battles in situations where disputes cannot be resolved by dispute resolution Boards (DRB's). This can be avoided by placing significant emphasis on establishing good client relationships prior to the bidding cycle that then continues throughout the construction process and into operations.

18.2   Relationship of Success to Risk Management

Success derives from well-planned and executed **risk management**.

Determining **success** can only derive from a common definition of the term.

What are the definitions of risk (management) and success? Using Wikipedia and Webster's Dictionary as the sources, they are:

'**Risk management** is the identification, assessment, and prioritization of risks (defined in ISO 31000 as *the effect of uncertainty on objectives*) followed by coordinated and economical application of resources to minimize, monitor, and control the probability and/or impact of unfortunate events…'.

Wikipedia defines **success** as:

'The achievement of a **goal** – *the opposite of failure*'.

Webster's Dictionary defines it as:

'*The correct or **desired result** of an attempt*'.

With success as a goal, applying risk management as defined above, will lead to successful projects.

When a project is about to begin, whether using a P3 or a design/build delivery model, there is a sequence of activities that must occur in order to increase the likelihood of successful delivery. The sequence of activities is:

1) If one doesn't exist, establish a risk management matrix (a simplified example is shown in Table 18.1).
2) Review lessons learned from past similar projects – apply them.
3) Select the best team suited to the type of project being delivered (designers, advisors, core team members, and project managers).
4) Know the client – ideally from past experience – adjust approach to suit.
5) Establish a regular team member communication protocol and follow it.
6) Establish a single point of contact with the client in order to maintain clear communication.
7) Establish a PMP – this entails setting out roles and responsibilities, team organisation, and reporting relationships, and communicating it to all team members, getting full team 'buy-in' and acceptance.
8) Establish clear leadership and control of the team reflecting it in the PMP.
9) Begin work!

Interesting to note that items 1–8 are undertaken before any project specific construction work has begun. Building the proper foundation is no less important figuratively than it is literally in creating a physical building foundation, or substructure.

Well-managed project risks lead to project success. The process begins with identifying categories of risk and planning of methods to deal with these risks. In general terms, once risks are identified, methods to deal with them need to be put into place. If risks are not identified and managed they could potentially lead to any combination of the following:

1) Project schedule not being met – potential for liquidated damages being applied.
2) Project being over budget – for fixed price contracts the contractor generally absorbs the over-run where there are no remedies for compensation.
3) Termination by the owner for a contractor default – failure to perform the work as specified.
4) Liability for third party damages – subcontractors, third parties.
5) Liens being placed against the work.
6) Counterparty law-suits for damages.

The following is a list of some of these key risk management methods:

1) Schedule planning and management – ensuring enough schedule float has been provided; scheduling to accommodate unknown permitting time frames.
2) Insurance – certain risks can be insured against; business interruption liability with defined caps in exposure.

3) Specification review – prior to contract award, plan to review and revise specifications with owner to ensure a deliverable project.
4) Personnel – assigning the correct number, experience level, and organisational structure to successfully manage the project; for example in urban areas, assigning a single person or team to deal with utilities locates and relocates.
5) Contract with owner – when the opportunity exists prior to contract award, review unfavourable or poor value for money (for the owner) contract terms in an attempt to correct.

There are at least four categories within which projects can measure success and therefore risks managed successfully. There are varying degrees of success derived from risk management, but ultimately a highly successful project achieves all project goals. In order to establish a framework for showing how to achieve success, it is worth reviewing at least four key areas. Achieving goals and successful risk management can be expressed in the following categories:

1) A project where no one has been hurt.
2) A project that has achieved all of its financial targets.
3) A project that has fostered and maintained personal and corporate relationships.
4) A project that has met or exceeded its technical requirements.

The four categories within which to measure success and to employ risk management methods are therefore:

1) Safety:
    a) All stakeholders buying into and following safety protocols;
    b) No recordable incidents;
    c) All participants empowered to work safely (prime and subcontractors, owners, consultants, etc.).
2) Finance or economics:
    a) 'On time' and 'on budget' delivery.
    b) Acceptable returns on investment (ROI) for all participants.
    c) Minimal claims.
    d) Sub-contractors and supplier ROI achieved.
3) Relationships:
    a) Happy client.
    b) Dispute resolution managed fairly and properly.
    c) Good on-going relationships with various authorities providing approvals and permits.
    d) Happy partners, designers, and internal team members.
4) Technical:
    a) Clients (users) programme requirements met.
    b) Benchmark technical specifications met or exceeded.
    c) Efficient and economical design.

Risk management techniques should be designed and used to ensure success in the above four areas.

These are the key measures of success, but how do you go about the process of winning the project and delivering success? What process do you need to follow in order to be able to undertake a project that can then be measured as successful or not?

18.3 Planning for Success and Managing Risks

Success is not the result of luck. It is the result of careful planning. 'Plan the plan' and 'plan to plan' are keys to success and the fundamental basis for risk management. A famous golfer once said that 'the more I practice, the luckier I get'. The same idiom applies, although the 'practice' component in the context of a construction project would be preparation and planning. I believe that 50% of the contribution to the success of a project occurs before anyone actually gets on a construction site. The same level of effort should be expended in selecting projects as in building them.

There are several key planning activities that should be used in preparation for creating a successful project. These activities should take the form of a risk mitigation matrix where possible. Filling out the risk matrix should then be one of the evaluation tools used in deciding whether or not to pursue a project. Keeping in mind the categories for success measurement, the planning activities (or risk measuring categories) are as follows.

18.4 Go/No-Go Stage

1) Project type:
   a) Do we have experience of this type of project; have we successfully constructed this project type in the past?
   b) Are there particular technical issues with the project type, and do we know what they are and how to mitigate them?
   c) Do we have a partner(s) that do have the specific experience required, with corporate philosophies aligned?
2) Personnel:
   a) Do we have personnel with specific successful experience in the project type?
   b) Can we commit senior project personnel to the project for the project duration?
   c) Do we know the subtrades and have relationships with them in the local project area?
3) Processes and procedures:
   a) Do we have proper internal and external communications skills and procedures in place and known to all parties?
   b) Do we prepare and follow a comprehensive PMP, and do we communicate this to all team members including the client?
   c) Do we have internal safety committees and clearly defined work procedures and processes?
   d) Do we provide the proper safety training both on and off site?
4) Authorities:
   a) Do we have experience dealing with local authorities and permitting agencies (building, utilities, planning, traffic, rail, etc.)?
   b) Do we fully understand local approval processes and timetables?
   c) Do we fully understand the selection process – lowest price, best value, and highest technical score?
5) Technical:
   a) Are the consultants the best for the project at hand?

b) Have we worked with them in the past?

c) Do we fully understand all liabilities and is insurance available for them?

6) Competition:

   a) Do we know the competition?

   b) What is their financial strength?

   c) Do they have bonding capacity?

   d) Who are their advisors?

   e) What is their track record with the client?

In designing the matrix, key evaluation criteria leading to a 'go or no go' decision should occupy the vertical axis. The horizontal axis should contain qualitative categories and mitigation actions that when filled in properly, will result in a risk management plan. In fact, a project will ideally have multiple levels of risk matrix, beginning with a 'go/no go' analysis or matrix. Beyond this category, matrices can be prepared for all aspects of the project as follows.

1) Specifications:

   a) General conditions

   b) Division content – vague or clearly defined

   c) Contradictions between sections

   d) Technically correct.

2) Drawings:

   a) Clear and complete

   b) Drawing and section references – complete and accurate

   c) Explanatory

   d) Technically correct.

3) Project mobilisation:

   a) Site access

   b) Permits

   c) Equipment availability

   d) Long delivery item ordering and costing.

4) Consultants:

   a) Experience

   b) Integrated into the construction team

   c) Previous experience working together.

Table 18.1 is a very generic version of a matrix that would contain many more categories and be significantly more comprehensive at the 'go/no go' stage of the risk mitigation process.

Table 18.2 shows a generic risk matrix based on the assumption the project is proceeding. Once again, it is a summarised version of what would likely be multiple matrices that would relate more specifically to various components of the project.

Keep in mind that when you Google 'risk management in construction' you get over 64 million results. Many lengthy books have been written on the subject, so what is contained here is only meant to point out some of the key components of risk management and to give brief guidance and direction as one investigates further the process of risk management.

Table 18.1 An example of some of the components of an initial 'go/no go' risk matrix.

Risk management matrix	'Go/no go' stage						
Category	Potential risk	Experience level	Technical issues	Available partners and staff	Safety record on similar	Do we know the authorities?	Who are the competitors?
Project type							
Personnel							
Process and procedures							
Authorities							
Technical							
Competition							

Table 18.2 An example of a generic risk matrix based on the assumption the project is proceeding.

Risk management matrix							
Category	Potential risk	Schedule mitigation	Price	Insure	Push back to client	Combination of measures – other	Risk not acceptable – do not proceed
Specifications	General Conditions vague						
	Section contradictions						
Drawings	Alternates acceptable						
Mobilisation	Granular location						
	Job site lease						
	Permitting process						
	Long delivery Items						
Consultants	Fee approach						

18.5 Summary

The following summarises the key components of success through risk management in North American projects:

1) Success is achieved through detailed risk management.

2) Risk management matrices should be used as follows:
   a) Initial go/no-go decision making
   b) Bidding phase
   c) Construction phase
   d) Post construction – warranty phase.
3) Risk management plans should form an integral part of a detailed PMP.
4) The Plan and the risk management process should be explained and understood by all team members including the client.
5) Rapid and effective communication of issues throughout the construction cycle (design through post construction) is critical to managing and mitigating risks.
6) Effective analysis of the mitigation measures to be employed should take into account the potential additional cost that could be simply applied, versus other less costly (but effective) approaches.
7) Every person involved in a project has a responsibility to identify, communicate, and manage risks in the same way that they should be empowered to do for safety matters.
8) **Regardless of the level of technical sophistication built in to risk analysis, the process is applied by humans. Making sure that project participants are 'invested' in delivering its success will reduce risks, or ensure they are mitigated. Often, a portion of team member remuneration tied to successful risk mitigation can be effective.**
9) Another effective method of managing risk derives from a pre-construction partnering session, followed by at least annual sessions during the course of construction. Again, this approach focuses on the interpersonal relationships of 'partners' (client and design build team) establishing direct lines of communication and a hierarchy of issue resolution based on each party's project objectives.

18.6   Recent Projects: A Success and a Failure

18.6.1   Project A – A Highly Successful Outcome for all Stakeholders – Southwest Calgary Ring Road, Calgary, Alberta, Canada

Project overview – 31 km of 6 and 8 lane divided highway, 14 interchanges, 47 bridges, one road flyover, one railway crossing (flyover), one culvert set, one tunnel, and three river crossings – 30 year operations and maintenance
    Delivery model – Public Private Partnership
    Developer – Kiewit (Canada) Development Company, Meridiam Infrastructure, Graham Capital, Ledcor Industries
    Constructors – Kiewit, Graham, Ledcor
    Client – Alberta Transportation
    Operations and Maintenance – AHSL
    Project Capital Value – CAD1.4 billion
    Initial challenge – large team with multiple parties in developer and constructor roles
    Project status – construction 30% complete

Keys to success:

- Clear PMP
- Heavy involvement of each company's senior management from the outset of the project
- Fully developed risk matrix identifying risks and mitigants
- Strong client relationships developed from the outset of the project
- Managed interaction between the design/builder and the Operations and Maintenance provider and developer

18.6.2 Project B – An Unsuccessful Project for Certain Parties Relative to Initial Expectations – Santa Clara County (California), Valley Medical Center

Project overview – 168 Bed Tertiary Care medical facility
  Delivery model – design/bid/build
  Owner/client – Santa Clara County, California
  Constructors – Turner Construction
  Project capital value – original value US$290 million (at tender award); revised value (after change orders delays, etc.) US$560 million
  Initial challenge – applying a delivery model not best suited for implementation with incomplete drawings and specifications and tender time
  Project status – in operation
  Lessons learned:

- On a large and complex project where the owner manages the architects and the builder builds the owners design, there is an inherent conflict between builder and architect.
- The project went to tender without a completed design so a proper assessment of the risks associated with tendering with an incomplete set of plans should have been carried out.
- The contractor was selected on the basis of the lowest price with incomplete plans leading to many extras, change orders, and schedule delays.
- This was not the appropriate delivery model for a large and complex project if design drawings were not 100% complete when tendering. Design/build or a Public Private Partnership would have been more appropriate, even though neither may have reduced the capital cost significantly, but they would have resulted in eliminating significant delays.

19

Early Warning Systems (EWSs), the Missing Link

Edward Moore, MICW BSc[1] and Tony Llewellyn[2]

[1] Chief Executive, ResoLex Ltd, UK
[2] Collaboration Director, ResoLex Ltd, UK

19.1   Introduction

How good is your organisation at managing behavioural risk on large projects? According to the statistics, the chances are that it is not very good. Despite the considerable resources that firms put into risk assessments, too many projects fail to achieve a successful outcome for reasons that were not anticipated, and yet in hindsight should have been predictable. As projects become larger and more complex, the challenge of getting your risk strategy right becomes critically important. So how do you improve your risk assessment system? Part of the answer lies in adapting your project systems to collect better early warning signals.[1]

Early Warning Systems (EWSs) have been recognised as a potentially valuable part of the risk manager's toolkit for many years, and there have been a number of scientific studies undertaken to understand how they work. This paper looks at the phenomenon of EWSs and why they are so important, not only to good project governance, but also to successful project outcomes.

Analyses of unsuccessful projects frequently show that the signs of impending disaster were available well in advance of the final failure. The hard facts that confirm a warning tend to arrive just before the crisis hits. There is, therefore, a very strong case for making better use of the data that an EWS can provide. So how do early warning signals fit within the traditional approach to project risk management? The answer is, not as easily as they should.

19.2   Look Outside of the Technical Bubble

Project teams often create a project 'bubble', within which they become almost solely focused on the technical challenges of design, programming, and delivery. The project risk register is created by asking the individual experts within the bubble to articulate the issues causing them particular concern. Consequently the risk register typically

---

1 For readers working in the UK construction industry, it is important to make a distinction between EWSs and the 'early warning' requirements set out in the NEC standard forms of contract.

Global Construction Success, First Edition. Charles O'Neil.
© 2019 John Wiley & Sons Ltd. Published 2019 by John Wiley & Sons Ltd.

comprises a schedule of issues representing either gaps in decision making or other 'known unknowns'.

Disruption, however, often comes from issues that appear from outside the technical bubble in the form of human interaction, (or the lack thereof). It is interesting to note how often a project post mortem points to the sources of distress that have been created by behavioural factors rather than by technical failure.

This near sighted approach to risk is compounded by a mind set that hangs on to the belief that projects can be planned in detail, and then controlled by the allocation of resources. This 'predict and control' mind set works adequately on projects that fit within a sequential programme where the patterns of activity are familiar. The approach becomes unstable as projects move from complicated to complex. In a complex environment there are too many potential variables to predict a precise outcome, as a change in one aspect of the environment may induce multiple potential side effects elsewhere on the programme.

We live in unpredictable times. The political, economic, and social factors that shape any major project are in a continual state of flux. Unpredictable environments require a new managerial skill set based around a philosophy of 'sense and react'. This requires a different attitude to risk, where the team learns to scan the horizon looking for signals indicating potential trouble ahead. Technical teams should be encouraged to learn to trust their own gut feelings, as well as listening more closely to the semi-articulated warnings that come from others. This is particularly true for the men and women who are working at the project 'coalface'. They are the people who are most absorbed in trying to deliver what others have planned, and are consequently aware of any tensions that are starting to arise.

19.3 Cultural Barriers

Based on the above observations, it should be a simple matter of good project governance to establish a process or system to collect early warning signals. Logic however is not enough. The studies that have been done around EWSs reveal that picking up the signals is only part of the task. Recognising the available data and then taking action on the information it provides is often a greater challenge to the project team. If the signs of impending disaster are frequently available to us, why do so many teams choose to ignore them? As is so often the case, it is partly down to poor systems, but more often is the result of human fallibility.

Humans are generally not very good at picking up early warning signals. In a review of the literature on the subject Terry Williams[2] and his colleagues pointed to three key areas where project leaders struggle:

1) A limited understanding of the full scope of project risk.
2) A low recognition of the implications of project complexity.
3) A lack of importance given to tacit/unwritten information shared between people, and how they react and respond to such communication.

---

2 Williams, T. et al. (2012) Identifying and Acting on Early Warning Signs in Complex Projects. *Project Management Journal*.

Table 19.1 Examples of behavioural blockages to the use of EWSs.

Limiting factor	Behavioural blockage
Organisational politics	The degree to which only certain information is deemed to be relevant. A preference for hard data over soft indicators.
Optimism bias	The tendency to believe that the problem will probably resolve itself and so no action needs to be taken.
Discomfort	Low desire to deal with interpersonal issues before they break into open conflict.
Shooting the messenger	Reluctance by junior staff to be the bearer of bad news.
Lack of faith in the response	A pervading belief within the wider team that no action will be taken and so identifying an issue is simply a waste of effort.
No external input	Rejection of observations from outside of the team/organisation on the grounds that they do not have the full picture.

The underlying message is that as projects become more complex, project managers and team leaders need to recognise that risk extends beyond the technical risk register. The last point around tacit information is particularly important.

Each of these three areas merits more discussion, but for the purposes of this chapter we will focus on the third point concerning the attention that needs to be paid to the human oriented social aspects of activity happening within and around the project 'bubble'. Many of the barriers to the effective use of EWS identified by Williams et al. can be seen to be cultural rather than organisational blockages. I have picked out a number of examples in Table 19.1.

These points illustrate a major challenge to the project leadership team and should ideally be discussed early in the project life cycle. Does your team believe that a process for identifying and collecting early warning signals would be valuable? If so, can you recognise the cultural and mental filters that will limit your ability to receive information and then act on it? Most rational leaders would say yes, confident in the belief that they would not be so blind to such potentially valuable data. It is important to recognise however, that the cultural and systemic forces that limit our thinking are strong and are often subconscious. Breaking through these limitations requires a degree of effort. This can be illustrated by our reactions to signals coming from 'gut feelings'.

19.4 Learning to Value 'Gut Feel'

Human beings are wired to seek out danger. In the modern workplace we are rarely concerned about physical safety. Our mental and physiological systems nevertheless remain programmed to watch and listen for signs of any threat to ourselves or our tribe. We often first pick up the sense of such threats in our stomach. Put simplistically, if we feel under threat, a part of the brain releases chemicals that can often be felt as contractions in our gut. Hence the phrase! Gut feelings can be stimulated by little more than watching the reactions or body language of another team member. You might notice a small reaction to a peripheral issue that feels too remote to be a real problem and yet is still bothering you. It can often be difficult to articulate these feelings. How do you write down a rational explanation based on a tightening in your stomach? And yet studies are clear that our gut instincts are often an accurate predictor of trouble ahead.

Of course everyone has days when they feel a bit pessimistic, and having a bad day is not necessarily an early warning signal. The value of having a large number of people involved in a major project is the ability to tap into the 'wisdom of crowds'. The trick is to design a process to pick up data that is quick to collect, easy to analyse, and that the contributors believe will be valuable. In Section 19.5 we set out some ideas for creating your own project specific EWS process. As pointed out above, however, the investment in building a system to collect this valuable data will be wasted if your team's mental filters reject the information before it has a chance to be considered. If, on the other hand, you have the foresight and perception to work through these cultural constraints, you will give yourself a much greater chance of managing your project through the maze of the complexity and uncertainty.

Everyone processes information differently. Perspectives on any issues will be determined by the individual's own personal window on the world. When a situation arises on a project each person acts on the information that they receive and responds in their own unique way based on their preconceptions, perceptions, and emotions. Their resulting behaviours can be positive or detrimental to the progress of the project, but are rarely considered as risks in themselves. We believe that they are a risk and therefore can, and should, be monitored and managed.

19.5 Case Study

A good example of a project benefiting from this approach is the Exemplary Low Carbon Project at the University of East Anglia (UEA). This project was a European Bank for Regional Development (EBRD) funded project and as well as providing a teaching and business incubator building for the use of the university it also provided a project where innovative processes and materials could be used as an exemplar in the modern low carbon built environment.

Having decided to use collaborative working as the model for delivery the UEA also embraced the principles of a neutral monitoring role on their project. The first step was to establish what exactly they could monitor. A workshop was held with the project stakeholders that allowed them to explore 'what success would look like' for all different parts of the delivery team. Once this was established and agreed by all the different composite parts of the team it was possible to build a delivery plan that would allow each project team member to achieve their personal and corporate goals whilst contributing to overall project success.

Once established, this 'modus operandi' for the project was turned into a project charter for the team to physically sign up to. This set the blueprint for establishing what behaviours would be needed for the charter to really work and therefore what needed to be monitored on the project.

The areas identified were:

- Whether the perceived final outcome of the project (at the time of asking) would achieve or exceed stakeholder expectations for the project.
- Was the project progressing in a way that would deliver the project aspirations?

- Was communication conducted in an open and honest manner?
- Were individual contributions communicated to and listened to by the senior management team (SMT)?
- Were the innovative elements of the project (processes and materials) capable of being commercially replicated on other projects?
- What was the level of collaboration between team members?
- Was the delivery team proactively identifying and mitigating risks?

The project team agreed that if all of these were being answered positively then the project would be delivered successfully across the board.

The ResoLex RADAR system (www.resolex.com) was used as the communication portal to deliver monthly feedback to both the team leadership and the extended stakeholder base and complete the communication loop by asking a series of questions on the areas mentioned above as well as some of the technical risks from the project risk register.

The results of the monthly project evaluation exercises were analysed by a panel of industry experts who created a report that was delivered back to the project leadership team. The SMT received a monthly report that identified how all the project participants perceived the project was being delivered against the desired objectives and behaviours, as set out in the charter, and what, if any, risks there were concerns about.

> Our project is a cutting edge exemplar project in both form and approach and in keeping with this we embraced an innovative approach to communication. We find that the RADAR report provided us with real insight into our scheme, just as we hoped. We felt that we might miss this information through only using traditional methods. The reports have added value by enabling us to tackle issues early, reducing conflict, and ultimately helping save time and money.
> –John French, Project Director, University of East Anglia

The benefits to the project identified by the project director above gave the project the communication loop needed to generate engagement with the project stakeholders. The additional feedback gave clarity to the risk management process and confidence that not only were traditional technical risks being identified and mitigated but that the project had a handle on the human dynamics of the project and the risks associated with them.

The premise of the monitoring and management mentioned in the case study above is that the information and knowledge that the project needs to achieve for successful project delivery is contained within the delivery team, but the challenge is gaining access to it, understanding it in the context of the project, and ensuring that the ever-limited resources are deployed on the right issue at the right time.

The two key areas to identify are:

1) Specific risks that are either new to the project or the delivery team feel are not currently being managed or mitigated effectively.
2) The divergence of perceptions between the delivery teams about a known risk or the general project progress.

Horizon scanning for new risks or ineffective management or mitigation enables this information to be communicated across the project delivery team so that all parties are aware of the risks and their role in managing or mitigating them. The elimination of surprise provides for a clearer and more strategic approach to risk on the project. The divergence of perceptions between teams is a more difficult risk to quantify; after all it is just personal opinion, some of them may in fact be wrong, so why go to the bother of understanding it?

As we touched on earlier in this chapter, it is an individual's perception, or opinion, that drives their behaviour on the project. Therefore if key individuals or different members of the delivery team have diverging opinions about issue X they are unlikely to have a coherent approach to resolving or managing it. From our experience, this is the seed of a dispute.

Most of those working in construction are acutely aware of the moment on a project where a manageable issue turns into a dispute. The red mist descends, and personal and professional egos stand between swift consensual resolution and full blown dispute. There is, however, a moment of opportunity when a developing issue can be worked through without substantially affecting the project, provided that the problem is identified and dealt with early. It is at this moment that the delivery of professional dispute services offer almost incalculable value to clients and the issue can be resolved in tandem with progression of the project.

When articulating the value of effective horizon scanning on a project it is important to consider the real cost of escalating disputes. By taking the results of a horizon scanning exercise and effectively communicating them across the project delivery team you can increase engagement and actually decrease the likelihood of issues escalating into disputes or unmanaged risks damaging the project. The transparency of information and knowledge on the project also enables the ever-limited project resources to be effectively targeted at areas of the project most in need at the most appropriate time.

19.6   Summary

Our purpose in writing this chapter was to engage the project community in a discussion on the value of early warning signals. As clients, project managers, and contractors become more familiar with the challenges of complexity, we envisage a shift in risk management thinking where spotting early warning signals is a standard part of every project manager's toolkit. The more people that we can help experience the power of this simple but powerful process, the sooner this aspiration will become a reality.

As with so many aspects of project delivery, however, the real challenge is to train our minds to recognise our own reactions to the signals we receive, and to accept the weak signal reported by others in or around the team. We do not underestimate the challenge this poses for many team leaders and project managers. Professional training encourages a focus on hard facts. As we progress into the twenty-first century, however, the volatile, unpredictable, complex, and ambiguous nature of the project environment requires a broader perspective. The ability to use 'soft' data to inform decision making may be the difference between project success and failure.

20

Construction Risk Management – Technology to Manage Risk (ConTech)

Rob Horne

International Construction Lawyer, Arbitrator and Adjudicator, Partner at Osborne Clarke, London, UK

20.1   Introduction to Technology in Construction

Construction, as an industry or business sector, is naturally conservative, or at least it is in the modern era.[1] The speed at which it identifies new opportunity, understands it, incorporates it and thus derives benefit from it could, perhaps, be termed glacial.

There is no doubt that many of the themes in this book around risk management, managing slim margins on high turnover and risk sit behind that conservatism. Perhaps, when one looks at major collapses, most recently Carillion, one would hope for even more conservatism in approach.

Watchwords in the industry in recent years have been 'evolution not revolution'[2] with a focus on prudent management of the status quo rather than radical rethinking of approaches. Underscored by the Farmer Report[3] there is a need to do more. The rather shocking, and no doubt deliberately so, strapline to that report of *Modernise or Die* should not be taken lightly. Nor should it be limited to the trades-based skill shortage that was the primary focus of that report.

The overall approach of *Modernise or Die* is going to be key to the industry and its major players if the technology available is to be identified and utilised to enable improvements in efficiency and margin. There should be no doubt that agile start-ups, focused on technology-enhanced delivery, will act as disruptors to the construction industry in a similar way to organisations in other sectors.[4] The only question is how far up the supply chain they can get; will they ever be in a position to oust the incumbents at the top of the turnover lists? Maybe not, at least in the sense that the tier-one contractors are currently seen. However, in terms of who is driving the market, and who are the 'go to' entities for clients looking to procure major projects, this is a much more difficult question to answer. The innovative companies will take the cream of the

---

1 Perhaps less so if one were to travel back in time to Brunel-driven engineering feats in the Victorian period.
2 This was, for example, used as the introduction line to NEC4 launched in June 2017.
3 The Construction Leadership Council. 2016. *The Farmer Review of the UK Construction Labour Model – Modernise or Die: time to decide the industry's future.* www.constructionleadershipcouncil.co.uk/wp-content/uploads/2016/10/Farmer-Review.pdf.
4 Examples are many, such as digital photography's impact on Kodak, the impact of Netflix and digital download of movies on Blockbuster, and so on.

Global Construction Success, First Edition. Charles O'Neil.
© 2019 John Wiley & Sons Ltd. Published 2019 by John Wiley & Sons Ltd.

work while others remain fighting for volume and turnover in the high stakes game of high-risk low-margin contracting.

20.2 What Do We Mean by ConTech?

ConTech, or construction technology, is a simple grouping of those technologies relevant to the delivery of construction projects. There is a huge diversity of technology in the market today of direct relevance to the construction industry, and the built environment more generally. Some are obvious and often discussed, others are identifiable, but perhaps have yet to be fully explored, and then there are technologies that construction has not yet found a use for.

However, underlying this, and when starting any discussion of what ConTech is, there has to be some context. At the time of writing, in the first half of 2018, there is significant discussion on just how fast ConTech will develop and grow. Where are we on that growth curve and what shape does that curve take[5]? In any event, there is no doubt that the context for any consideration of ConTech is rapidly changing and therefore principles have to be the primary consideration rather than details.

To provide some context, some examples of the types of ConTech now being used, to a greater or lesser extent, and those technologies with something to add to construction but not yet really utilised, are set out below.

20.2.1 The Currently Used ConTech

Building information modelling [6] *(BIM).* This is one of the technologies that has actually been used in the construction industry for some time. While used, it is under-utilised. The principle behind BIM is to bring all design information[7] together into a single model that captures all building data and changes as the construction proceeds. This enables efficient construction through avoiding clashes and consolidating all knowledge and input to a single information point. However, difficulties arise when you try to classify BIM in anything other than the most generic sense.[8] The principle, however, is that the BIM model incorporates all data for a project as a single point repository. How you get everything into the model, and indeed what is the 'everything' you include, remain as areas for development.[9]

3D printing and off-site manufacture. 3D printing is one example of the use of robotics and can take many forms. The essence is that it is an automated process of building

---

5 Explained further below.
6 For present purposes COBie is treated as a subset or specific type of BIM.
7 Arguably a BIM model can incorporate and host more than just the design; it can include cost data, progress information, etc.
8 See The Winfield Rock Report, which asked what does BIM Level 2 actually mean, and did not receive two answers that were the same from all those who replied. 2018. M. Winfield and S. Rock. *The Winfield Rock Report: Overcoming the Legal and Contractual Barriers of BIM* http://www.ukbimalliance.org/media/1185/the_winfield_rock_report.pdf.
9 For further reading see *One More Level v Another Dimension* by Rob Horne first published in the ICES Journal April 2016.

something. This has generally been used on a relatively small scale up to now,[10] in particular looking at intricate components or design less, or not at all, constrained by traditional building techniques.[11] However, there are now more examples of this technique being used on larger scale items, including the automated construction of simple houses.[12] In its current state of smaller component manufacture it is an important element, potentially, for off-site prefabrication. Off-site manufacture is a means of deconstructing a project so that it can be manufactured in a controlled environment off-site and then, relatively simply, built again on-site. The intention being to create production line-like facilities, similar to the automotive industry, where quality can be more easily checked and controlled, and economies of scale employed.

Drones. More properly defined as UAVs[13] these are remote vehicles that can be used for a wide variety of tasks, from investigation and monitoring, to record creation and progress tracking.[14] They can be fitted with a range of monitoring and survey devices to capture information about the project and its environment from initial site mapping (surface profile and subsurface to an extent) to maintenance surveys during the operational phase of the asset. In addition, although of lower impact to construction, drones can be used to make deliveries of products and materials. The size and scale of most construction, however, makes this use of only minor relevance.

Enhanced/augmented reality. This type of technology allows the users to 'see' the end result or any stage of the construction practice as it might look as a graphic image. There are many different types of visualisation techniques, from simple wire models and static fly, through to images of more complex 3D models that can be seen in situ on site, or 4D models that follow construction methodology to show progression of the build. The current usage is more at the basic end of this scale, with some pioneer projects using more advanced systems.

20.2.2 Less Used ConTech

Robotics and automated vehicles. The principle here is slightly different to 3D printing and captures those construction processes that would normally require a human interface but that can now be delivered by computer control. There are varying levels of automation or robotic control, ranging from relatively simple remote control, like the majority of drones, to an intermediate stage of partial automation where a driver remains but part of the operation is guided by and completed via an automated process. Finally there is full automation where the vehicle can move from one tasks to the next with no human intervention. Looking at specific examples, this type of ConTech would include automated bricklayers, automated graders or automated painters. There is also something of a crossover between a drone and an autonomous

---

10 The largest 'structure' produced by 3D printing as at May 2018 is a 12 m long bridge for use over a canal in Amsterdam (www.turing.ac.uk/media/news/researchers-working-worlds-largest-3d-printed-steel-bridge-celebrate-another-landmark).

11 The discussion around 3D printing is generally focused more on aesthetics and engineering opportunity rather than commercial opportunity, so while undoubtedly an important part of ConTech it is not discussed at length here.

12 www.theengineer.co.uk/europes-first-3d-printed-house.

13 Unmanned aerial vehicles.

14 See for example the use of drone footage for the development of the ITER Reactor in France https://www.iter.org/news/videos.

vehicle designed to survey and measure the interiors of buildings to assess progress, etc.[15]

Materials tracking and wearable technology. This technology is intended to allow better knowledge and therefore management of equipment, material, and labour on a project. This type of technology can take many different forms, from relatively basic 'just in time' materials ordering systems simplifying and automating logistics control, through to site clothing capable of monitoring the location and life signs of people on site to ensure they are working safely and for appropriate periods of time.

Digitised documentation and intelligent analytics. As a broad catch-all this is the use of computer-based documentation to manage the commercial processes running in a parallel track to BIM for the design process. Despite the relative availability of digital documentation there is still huge inertia in the construction industry towards hard copy or 'simple' soft copy[16] rather than digitised documents. While that approach has changed substantially in design, commercial processes and on-site delivery are generally far behind. Linked to digitised documentation is what you do with them, which is the area of intelligent analytics. Intelligent analytics systems provide feedback on risks being generated and resolved across a wide range of processes. The maximum benefit from intelligent analytics comes from it scanning across the business as a whole and filtering through market norms so that those directing the business can understand very quickly what risks they are carrying and what return they are making on those risks without any subjective optimism bias being layered into the reporting process.

20.2.3   Not Yet Substantially Used ConTech

AI. Artificial intelligence could have many applications in the construction industry, from automating commercial process, such as the production of the contract, through to producing design and specifications without human involvement. The primary benefit of AI is that it is a learning system so it can improve both its commercial awareness and risk identification by drawing on multiple sources of information very quickly to assess and present options. It could also rapidly consider the whole life implications of solution alternatives, predicting maintenance cycles and ease/cost of maintenance. The downside of true AI or deep learning systems is that, unlike normal computer programmes, the results are no longer easily predictable. The AI becomes something of a 'black box' of information. Additionally, to get the most out of AI requires a significant amount of training, and that training has to be done very carefully so as not to 'teach' bad behaviours, prejudices or biases.

Blockchain and smart contracts. This technology is widely used for electronic currency transactions.[17] In very simple terms, it is a means of tracking, verifying, and securing multiple transactions or processes. The idea is that those involved in the Blockchain are all involved in processing and approving each transaction or process completion, removing the need for a central directing authority or a single point of responsibility. Each transaction or process completed is electronically linked to the one before it so that a very clear and difficult to modify audit trail is compiled. One way to

---

15 https://www.doxel.ai.

16 'Simple' soft copy would be a pdf or similar, which, although it is electronic, does not achieve any of the benefits possible of full digitisation such as checking, tracking and integration with other documents.

17 For example Bitcoin is based on Blockchain.

envisage Blockchain in a construction sense is that it is very much like a construction programme with F–S logic[18] pinned up on the site office wall for everyone to see and mark up as work is completed. Smart contracts are often associated with Blockchain and are really a step on from digitised documentation and intelligent analytics where contracts are produced, issued, and carried out with no human intervention at all. These could be envisaged in the form of the supply of materials contracts for example.

20.3 ConTech as a Tool Not a Toy

The danger with any discussion relating to technology is that it becomes a discussion of what the technology can do[19] as a technology rather than how it can be harnessed to provide a focused solution to an existing or future problem. Equally, the focus of most discussions on technology in construction from tech providers is, unsurprisingly, about the technology itself and what it can do, often with little relation to how it could be used or how one could transition to using it. Other frequent discussions are with lawyers who focus on either the new risks being generated or the legislation needed to regulate the new technology. Again, this is not surprising, but it is unlikely to move the conversation forward in relation to the use of ConTech for commercial functions as the risks and regulation is often out of context when looking at how it could be used and is stripped away from the balancing explanation of the benefit that could be generated.

The conversation needs to change, radically, if the opportunities presented by ConTech are to be utilised within the commercial environment. However, that does not mean that ConTech should be embraced, from a commercial perspective, without challenge. The question, particularly from a commercial perspective, to ask in relation to any and all technology is 'so what?'. The risk–opportunity balance is as important to the consideration of the adoption of ConTech, as in any other area. From a commercial perspective that sceptical eye is perhaps even more important as the role of the commercial function is to ensure profitable delivery of projects. However, that profitability is not a static arena, more so now than ever before. Retaining or improving margin requires better processes and ConTech has a role to play in this.

A few of the macro issues are set out below; the reasons why there is reluctance to adopt ConTech in the commercial sphere are:

- ConTech is a distraction. The vast majority of companies involved in construction projects do not have, as part of their core business, developing and using new technology. Therefore, without having a clear focus on the output you to want to achieve through its use, technology will subtract from, rather than add to, effective and efficient project delivery. When working to a very tight margin everyone across the commercial delivery team must remain focused, there is simply no spare time for playing with these toys.
- ConTech does not provide commercial solutions. The technology available to the construction industry is focused on design and on-site delivery from BIM to augmented reality and that has no place in commercial processes that focus on

---

18 Finish to Start logic so that the next activity cannot begin until the predecessor is complete, e.g. build walls (F–S) build roof.

19 I will return to this in Section 20.8, referenced as the 'Terminator' effect.

contractual operation. Intelligent analytics and digitised documents cannot replicate or replace the human interaction necessary in the commercial operations sphere; commercial operations are driven by people and human interactions in which trust is a large component and ConTech cannot provide that personal assurance.

- ConTech R&D is expensive. Contractors work to very tight margins and are in something of a 'catch 22' situation. The way to improve margins is to become more efficient, but to become more efficient requires time and investment, and that time and investment needs to be developed out of the current profit margin. A necessary consequence of R&D is therefore a drop in profitability, but where and how far off is the return on that investment? If investment is necessary just to retain the status quo in terms of market position I am better off taking my profit now while others develop a new system that I can adopt later. Further, any savings a contractor makes are in danger of simply being taken by the developer/client/employer so that even with the extra effort the net result is that the saving goes to someone other than the one investing in the R&D.

- ConTech is unproven and unreliable. Adopting ConTech means exposing the company to additional and different risks that it is not used to dealing with. Integrating any ConTech with current systems is hard and may not only lead to a failure of the ConTech element but may have cascade consequences into other areas of commercial control, undermining the whole business strategy. A contractor simply does not have the margin in any work it is delivering to deal with the consequences of any failure of the ConTech; the supply chain won't bear the risk, designers and professionals exclude the risk, and employers ensure that contractually all the risk stays with the contractor. Any changes therefore need to be very small and must be slowly adopted and incorporated to match existing systems so that when things go wrong the problem can be identified and fixed before any damage is done.

- Retaining control. The commercial function is at the very heart of the business and releasing control of that, or any part of it, to a machine means releasing control of a core part of the business. Decisions could be being made within the ConTech that do not align with the overall business strategy and once started can be difficult to unwind. Using ConTech without human influence and control cannot be done in the commercial sphere because the decision making is too nuanced and requires the consideration of too many variables to give any control away; as a means of providing some organised data and records it may have a place but it cannot grow beyond that.

Before ConTech can take a substantial role in any organisation all of the above resistance (or inertia) points need to be dealt with and answered; not by the technology team but by the commercial teams themselves. The 'so what?' questioning of ConTech should deal with most of these but only if behind the 'so what?' is a genuine interest in change and improvement. The 'so what?' needs to be focused on what is the best change[20] to make rather than whether to make any change at all.

The days in which it was perceived as acceptable, or something of a joke, that anyone, up to and including the chief executive, could not use their own email or the internet must be consigned to history. The technology prejudice must be set aside and seen as just as unacceptable to a business as other forms of prejudice.

---

20 The best change could be driven by the one that will have most significant business impact, the one that is simplest to adopt, the one that is cheapest to adopt, or even simply the one that is likely to be utilised rather than viewed as a toy or 'something for the youngsters'.

ConTech is no more a toy than the first mechanical diggers were to the workers with shovels or the first CAD systems were to designers used to using paper and pencil. ConTech is a tool that can be utilised, or not, at the choice and risk of the industry as a whole and the individual delivery companies within it. However, understanding the 'tool rather than toy' approach to ConTech is difficult in the abstract, it needs context and grounding. Suggestions for that grounding are in the Section 20.4, looking at the internal mechanics of major projects and how similar they are to developments in smart cities (including what a major project set up could learn from a smart city mentality). Later in this chapter is a description of a project utilising a connected set of ConTech and what it might look like (Figure 20.2).

Bringing scepticism and challenge to the adoption of ConTech is good and healthy but those challenges should not become a blindfold to where ConTech could add value. Asking 'so what' to drone technology regarding mapping of a site is good. However, taking that as an example, the benefit of drone mapping is that it can be completed faster and more cheaply than traditional methods for achieving a reasonable to high degree of accuracy. Adding in ground penetrating radar surveys from the drone as well as laser levelling can give a detailed enough starting point that the risks and opportunities for a project can be understood from a very early stage. As the surveying can be done more quickly and more cheaply there is a positive incentive to get this done at tender stage so that cost and risk profiles can be more robust and built on real data rather than a guess or historic surveys that could be out of date, and/or come with no warranty for accuracy. The 'so what' question is then answered 'because it can de-risk the project and add to project control and understanding as or before a tender is submitted'. The same level of challenge needs to be applied to all ConTech.

20.4 Major Projects – Temporary Smart Cities

Projects have grown in size[21] and as they grow the logistics and infrastructure required to enable delivery , on a temporary basis, become more complex and extensive. The supply chain itself is diverse and the risk pattern within it perhaps as unpredictable as the weather. The integration of the supply chain and logistics generally is in the process of a step change towards being considered a substantial and substantive project in its own right.

So far, the construction industry is still chasing the idea of a co-located team as the ideal tool for commercial management. However technology, as it currently exists, can allow a far deeper and more meaningful integration. Through technology we can move away from integration of the client and tier-one contractor to full integration of the entire project team. The signs and shoots of understanding of this can be seen in BIM but that is just a single element of the potential of integration.

The project delivery team will be living in a temporary smart city, a hub for managing the translation of a digital project into a real world one. It has the sole purpose of

---

21 Current mega-projects include CERN and ITER in France, Hinckley Point and HS2 in the UK, the delivery of the football World Cup in Qatar, train and metro lines leading to whole new cities in the Kingdom of Saudi Arabia, water projects (3 Gorges Dam and North–South Water Transfer) in China, California high speed rail in the USA, etc.

supporting the delivery of a project, but what does that mean, and why is it relevant to commercial management of projects?

20.5 Smart City Principles

What then is a smart city and how does it has relevance to the logistics hub of a major project? A few of the key principles, directly relevant to a construction project, are set out below. At a headline level, however, there is a general correlation between smart city principles such as interconnectivity, record and data collections, and optimisation of energy usage and the principles that will or should govern the smooth and effective governance of a construction project. These are taken in turn as follows.

20.5.1 Interconnectivity

The first guiding principle of smart cities is that they should operate as a single system with each part of the smart city integrated with, available to and sharing data with all the other systems within the smart city. The information flow should be unhampered and as close to instantaneous as possible with all systems 'talking' to each other.

In a smart city context, this could relate to management of traffic, so where traffic is slow it is automatically re-routed to a more efficient route because the traffic management talks directly to the traffic cameras and anyone in the city can have their sat-nav system join the smart city network.

How is that relevant to a major project? The beginnings of this type of interconnectivity exist already in colocation and the use of joint storage and prefabrication areas for the various delivery partners to the project. However, growing and deepening that theory is possible. Rather than tracking orders separately from other site operations, particularly for long lead items, the tracking (both during manufacture and delivery) should integrate to the systems so that the project team can see exactly when delivery will be made and compare it to progress on site. Where an area of the site is being influenced by an unexpected event,[22] plant and labour could be automatically re-directed to an available area. Site photographs taken and records entered are immediately cross checked and verified against each other and against outside data sources[23] and anomalies flagged for review and action. Once there is interconnectivity at a sufficient scale the opportunities grow rapidly, but it takes a leap of faith to move away from conservative, contractual, and commercial models with each party hiding behind privity of contract and trying to manage their own delivery problems in isolation.

20.5.2 Records and Data

The underlying mechanism that allows a smart city to function is intelligent and comprehensive data collection. This would include tracking of energy usage across time and in different locations within the city, and tracking and recording movement and footfall,

---

22 This could perhaps be a problem on the ground or the failure of a statutory undertaker to complete its works.
23 For example, where poor weather is recorded it is immediately cross-checked against independent weather measures.

both to understand how the space in the smart city is being used but also to generate patterns for maintenance, where high impact advertising should be placed, when and where accidents are more likely, etc.

The collection and management of data is potentially even more important to a construction project that is growing and delivering a digital project to a real world asset. Of course for a construction project the data collection does not begin or end with the physical construction on site but overlaps either end of the phase. The data collection will start as early as possible, including negative decisions taken,[24] in the project concept phase, through to design and construction and on into the operational life of the project.[25] Equally, data collection and analysis shouldn't be limited to the physical construction works but also how the site space is used and how, when, and where power is accessed.

20.5.3   Energy Use and Optimisation

Any smart city concept will start with an understanding of power needs, not just as a total consumable value, but on an hour by hour basis, with seasonal variations so that power can be obtained from the most effective source possible and available at the relevant time to allow delivery of what is needed, where it is needed, when it is needed. The smart city will be looking at multiple different methods of energy delivery from renewable to more traditional production and energy storage in addition.

A major project will have very substantial power consumption and that power consumption could be spread over a wide geographic area.[26] Managing the power delivery and consumption effectively, whilst ensuring power is always available so as not to slow down the project, is a major challenge. Obviously a project cannot set down the infrastructure for a smart city on a temporary basis but principles can be drawn, such as identification of efficient sources of power[27] and storage of locally sourced cheap power, until such time as it is either needed or can be offset against more expensive power available on-grid. Smart city principles of energy use and optimisation should be simpler for a project where the energy use, in terms of volume, time, and location, should be easier to predict and therefore provide for.

20.6   'Smart' Commercial Management

Smart city principles have something to offer the on-site logistics, and possibly delivery teams and post-construction in the operational phase, but how is that relevant to commercial management? How do smart city principles relate to managing time, cost, and quality of project delivery?

---

24 The decisions not to pursue certain options.

25 This would be the case whether the project is a tower block, a road or a development across a substantial area.

26 Take a project like HS2 in the UK, for example, which is 230 km in phase one and a further 530 km in phase two, in a thin ribbon of activity with some 'hubs' of activity around stations.

27 Can mobile solar generation be operated, is there an opportunity to use regenerative techniques (for example power deployed in a crane lift could be recaptured during lowering), can 'smart charging' be adopted for electric vehicles to take advantage of cheaper troughs in electricity demand.

For years the almost unanimous cry from consultants to projects looking at the commercial aspects was 'records, records, records, get more records, get better records'. Make sure you capture everything that happens. This was dealing with the evil of a lack of data and information. With the rise of email as a means of communication in particular, the need for more records has somewhat diminished on most major projects. The cry for more has been substantially replaced with the cry for better records and more precise documents; ones that really focus on the relevant issues.[28]

More and better records and data are always going to be relevant, but the key to their effectiveness is analytics and data processing. This is the key area where ConTech can and should provide the exponential improvements so often talked about. There is no value in 500 000 emails if that is basically 499 500 that add nothing and 1% that are of value. Finding the 1% will be time consuming, costly, and there is a high risk that some part of the important data set will be missed. The process is, in reality, extremely simple. There needs to be data capture followed by data analysis, then application to generate an outcome.

Data capture and analysis, however, should and can extend beyond the end product to also include the process of construction. That way the commercial team is enabled, through proper data and fast, accurate analysis, to make decisions about the project to increase efficiency. Which systems can be joined together to provide the quickest and most substantial benefits is obviously a starting point but the context should expand from a site to a region to the whole business in terms of connectivity.

20.7 Dehumanising Risk Management

In *Human Dynamics in Construction Risk Management*[29] one of the underlying principles explored that instead of the process it was the human dynamic and involvement in the commercial process that was the single largest driver of success or failure of a project. Is there then a way for that human dynamic, or the variability within the human dynamic at least, to be removed or reduced through the use of ConTech? If it can be then, and this is the million dollar question, how do you do it without losing control of the process and continuing to drive performance and improvement across all aspects of the project?

First, do you want to lose the human dynamic? In principle this seems a simple and straight forward question, but it is not. While removing human fallibility and subjectivity may give more consistent results, the human dynamic of a major project is important and is likely to remain so into the future. Having, understanding and implementing joint objectives is at the core of a successful project. This is particularly the case where there is change or where the original concepts are tested and challenged, possibly to beyond breaking point. Perhaps then the idea and intent is not the removal the human dynamic but the enabling of it to operate where it is appropriate and allow other process to take the lead where they can add more value.

---

28 This is always difficult because often, using traditional commercial management techniques, the key issues are not known until close to the end of the project, by which time the records are, of course, the records.
29 Written by Charles O'Neil and published in 2014.

Figure 20.1 The 'commercial' triangle.

How then do you take and implement the best elements of ConTech without completely removing the human dynamic?

A typical commercial model for a project is often related to a triangle as in Figure 20.1.

To enable, or realise the benefit of, dehumanising commercial and risk management, a solution needs to be identified to retain the balance within the triangle. It cannot become distorted so that additional risk is carried in any limb of the commercial matrix or that the risk balance between ConTech and the human dynamic becomes distorted in the model as a whole or any one aspect of it. That distortion will likely unbalance the system either by requiring more human involvement (thus removing efficiency and reducing integration) or creating bottle necks within the process, slowing down decision making rather than speeding it up.

ConTech is a disparate group of ideas and concepts at different stages of development and implementation. To try and capture the ideas and value in ConTech from a commercial perspective they are best broadly grouped together to provide solutions, rather than dealing with individual sources of ConTech. This keeps the focus very firmly on outcome drivers: what is the problem you are trying to solve and what is available to help with that?

There are many different methods of categorisation that could be applied but the following seem best suited to drawing out commercial uses of ConTech[30] as they move, generally, through the construction process.

20.7.1 Data Generation and Capture

This category will encompass drones, data tracking from labour, plant and machinery,[31] and other ConTech designed to track, measure or record information about a project. This is, in many ways, the simplest form of ConTech and at least one commercial application is readily apparent. The more information and detail one has the better 'armed' one is for any commercial analysis. Any ConTech solution, to derive commercial benefit, should be capable of cross referral and verification. The days of a site photograph with a

---

30 Commercial as opposed to design or on-site construction.
31 This would primarily be through RFID (radio frequency identification) and is often referred to as just RFID.

copy of the front page of a daily newspaper to validate it are gone. Not only do you need to know when the data was collected but also from where.[32]

If, however, you can have multiple data sources that cross verify each other automatically[33] the data set becomes a compelling tool and a source of highly persuasive information to be used in analysis, whether that analysis is automated or traditionally carried out. The validity of data, and therefore reliability, twinned with frequency, to generate an opportunity for commercial management should be the focus. That it can be generated relatively cheaply, once the systems are in place, is a bonus or incentive to actually do it. However, the generation and capture of large quantities of data can create new and unexpected risks.[34]

Looking back at the 'commercial triangle' the data collection ConTech can and should be applied equally to gathering information on progress whether that be from the perspective of quality, time or money. The balance is important here so that the information about any area of the project is captured, and can therefore be analysed, in the same way. One does not want a manually entered accounting spreadsheet twinned with drone based progress images and handwritten quality statements as it will distort the commercial balance. The excellent drone images cannot directly correspond to the quality statements or the accounts system so they will most likely to considered in isolation.

20.7.2 Data Analysis and Presentation

This group includes ConTech such as BIM,[35] augmented reality and intelligent analytics, i.e. those technologies focused on taking data sets and providing assessed outputs[36] and conclusions based on defined criteria to enhance the understanding of what is happening on the project to support decision making. At a simplistic level the commercial application here is again obvious. ConTech can take some, or indeed all, of the variability and subjectivity out of the analysis process to provide consistent comparative results.

Importantly, the data analysis tools available through ConTech should be kept separate from decision making tools in analysing their value and potential usage. Mixing data capture and data analysis can lead to a misunderstanding of how the system works and/or at what points you can reinsert the human dynamic. The BIM environment, for example, does not propose or implement solutions to any problems generated. What it does it bring together disparate data, presents it in a way that is easier for a human reviewer to understand, and applies its own decision making process. The BIM model could, for example, identify a clash within the design;[37] it would not however resolve that clash.

---

32 Although a picture can paint a thousand words, that picture can be highly misleading if, for example, it is taken with one's back to the current work front, back into open land not being progressed, or at a lunch time showing few or no workers on site.

33 For example, a drone confirming physical progress on a time lapse basis over a week, cross referenced to the labour, plant and material tracking to show efficiency, cross referenced to weather data, cross referenced to a cost ledger, cross referenced to site signing in/out records.

34 Imagine images captured by a drone of illegal activity on-site, or of health and safety breaches, or amounting to privacy invasion of the taking and storing of personal data relevant to the GDPR.

35 BIM could really be considered a transition technology straddling capture and analysis. It is included here in the analysis group as that is where the true benefits of BIM are derived.

36 The form the output takes will be dependent on what is being analysed and why.

37 A good example of a clash is ceiling void pipework running in line with a series of columns or a foundation trench crossing a drainage run.

The commercial application in this grouping should take the large volumes of data generated on modern projects and filter that information to a manageable human interface level, no doubt with a graphic based interface. The use of these data analysis techniques then makes it quicker to identify potential problems, and the sooner an issue can be identified, the more options will be available to solve it.

The real power in the analysis tools, however, is in the data set they draw from, and ensuring the correct analysis is applied to that data set to generate the useable information to enable commercial decision making. Joining the data analytics and presentation to the previous group of data generation and capture gives a solid grounding of information and tools to use that information. Very much as with data gathering ConTech, balancing the analysis across the 'commercial triangle' will enable consistency of analysis, which will lead to the efficiencies and benefits of the ConTech being realised. Without the consistency, anomalous results become more likely, which would lead to questioning the ConTech rather than the system and environment in which it is working.[38]

The ultimate use of this type of ConTech would be sampling from a data set wider than just the project in question in order to establish current market norms and baselines. Utilising the wider data pool would mean that the commercial process, using ConTech as a tool, could move away from being reactionary, based on data generated, to being predictive of similar issues arising in similar circumstances on other projects. We then come to the interface point with the last group of ConTech, the option generation processes.

20.7.3 Process Automation

Building on the data generation and capture, and the analysis tools above, the next group of processes are focused on automating commercial processes. The types of ConTech involved are machine generated progress reporting and cost reporting, as well as the introduction of smart contracts and ultimately automated payment mechanisms.[39] The key point from a commercial perspective here is taking some of the process based preparatory work away from the human workforce, enabling a lighter more specialist commercial team to have oversight of the project.

As with the analytics group of ConTech the viability and value of the process automation ConTech is derived from the veracity and extent of the original data set. If you want an automated cost reporting system then that system will work best if the data generation and collection ConTech is effectively gathering all information relevant to that cost report.[40]

The secondary or additional benefit that can be generated through automated reporting by ConTech is the removal of optimism bias and subjectivity in the reporting process. The system will respond and generate the report based on the available data (facts)

---

38 It is the system that is important, for example a digger with an undersized bucket and no propulsion would still work but would be highly inefficient, and an oversized bucket and insufficient counterbalance equally reduce efficiency, but when they are balanced you have the right tool for the job and output can be optimised.

39 See for example Hyperledger (https://www.hyperledger.org).

40 A cost report may only deal with cost expended but it could also deal with value generated to enable a fuller picture or an extreme EVA (earned value analysis) to be generated.

applied to a set of rules (the contract documents) without filter or interference. Of course, that reporting may not tell the whole story and there is a time and place for subjective review and interpretation. However, being clear on what that subjective interpretation is, and how many layers of optimism bias have been applied[41] before the final report is generated, is important for making informed commercial decisions. Again, it is important to distinguish the ConTech that can generate a report from that which might act on a report and alter any system of approach on the project.

20.7.4   Option Generation and Implementation

The previous categories of ConTech have all focused on providing the human commercial team members with information to enable them to make better, quicker, more informed decisions to drive the project forward. This final category is all about allowing machines into the decision making process itself and includes intelligent analytics and AI. This is where the focus of many ConTech articles and discussions rests, but is a jump to the end game without thinking about the steps along the way, and how long it could take to get to an automated system of commercial control.

The commercial benefit of this ConTech is very much like automated reporting, i.e. twofold. The first benefit is in the ability to reduce the size of the commercial team and the second is to remove, reduce or better identify and understand the subjectivity taking place in the decision making process. However, it should not be seen as an absolute handing over of control in a single step, the ConTech could and should be seen as any other member of the commercial team. One does not assign full commercial responsibility and decision making powers for a major, or even mid-size, project to a graduate QS. There are checks and balances in the human commercial world and so should be in the ConTech equivalent.

The first step might be using ConTech to generate options and opportunities based first on the data being produced on the project, but also drawing more widely across the business or even broad market intelligence and trends. The next step might be automation of some processes such as updating of an early warning or risk register, etc. so that there can be incremental growth and reliance on the available ConTech, but always challenged at each stage with 'so what?'.

There are then a few types of ConTech that fall outside the groupings above, the key one of which is blockchain. The basic principles of blockchain have been described in Section 20.2.3 and the specific application that blockchain could have is set out below. It could be that, rather than balancing areas of the 'commercial triangle', blockchain is the triangle, and the parameter the risks fall within.

20.8   Joining the Dots for Exponential Growth

Most conversations around ConTech start with a description of the technology or the technological solution to a perceived problem. The solution is considered, generally, by

---

41 The optimism could start with a manufacturer, which is then overlaid with further optimism from the supplier, with additional optimism layered on by a subcontractor and then the contractor finally adding its own optimism. Although each layer maybe relatively thin the cumulative effect can significantly distort the end report.

reference to a single source of ConTech operating to solve a very specific, and often contrived, factual scenario.

The trouble is, those looking at the adoption of ConTech are not 'technologists', they don't need to know how it works, in any degree of detail, just what it can deliver. Those same people very much live in the real world where, with almost the same certainty as death and taxes, one can say that construction projects change. If the ConTech being considered cannot adapt (and be demonstrated to be able to adapt) to those changes at least as well as a human then the resistance to change or inertia in the status quo will not be overcome. Of course, finding a single source of ConTech that can provide the adaptability and opportunity of a human interface is difficult if not impossible. However, that is like trying to find a better way to build a house by only looking at some bricks and some cement; it is not even a complete set of one operation let alone the holistic approach that will be needed to match up to a human commercial management approach.

To get the benefit from ConTech one cannot look at individual solutions in isolation; one must look at the overall issues and problems that one is trying to address and then look at how the solution is built from multiple components. Some of those components may be pure ConTech, others might be ConTech-supported human involvement, and some may have no ConTech application at all but be an adjustment in thinking or practice to enable the ConTech to operate effectively. The siloing of ConTech will inevitably limit and inhibit its functionality and ability to deliver results for the problems it is being applied to.

The question then very rapidly becomes 'well what ConTech join together?'. The answer to that, of course, is that it depends. It depends on what you are trying to achieve and how well integrated your systems can be with other members of the project delivery team. The commercial triangle described previously then becomes a tetrahedron, or a 3D triangle, to represent other parties to the project, all their interface points and the shared risk between them all, see Figure 20.2.

It is often said that ConTech is on an exponential growth curve. That may be true of the ConTech itself but is unlikely to ever be true of the process and usage that follows behind it, which is more linear in development growth. However, rapid and increasing growth can be achieved following the adoption of ConTech, and the more integration between different types of ConTech that can be achieved, the closer to that exponential acceleration the processes are likely to be. The key point being that iterative, evolutionary change in systems is most likely to give a straight line progression. That may leave

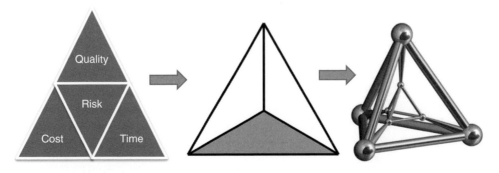

Figure 20.2 The progression - commercial triangle to tetrahedron to 3D triangle.

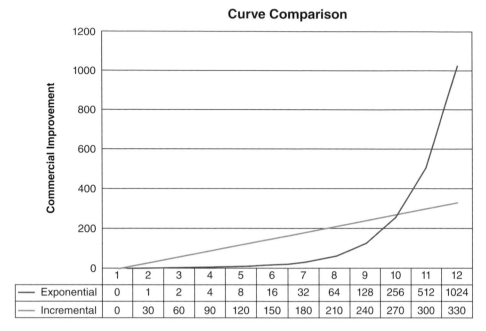

Figure 20.3 A very simple exponential versus straight line change.

your process 'artificially' ahead of the exponential curve in the early stages but will leave you well behind very rapidly.

The unknown is: where exactly on an exponential curve might we be now? The spend on R&D, if ConTech is in periods 1–6 shown in Figure 20.3, is unlikely to derive much value. By period 10 the curve has moved but if you did not have processes for dealing with ConTech already, trying to catch that curve as it accelerates up the page is going to prove extremely difficult. Anyone tracking that curve more closely will necessarily have a commercial advantage. The switch point between linear, incremental, evolutionary improvement, and development to the opportunities driven by ConTech is very difficult to spot, other than in retrospect.

Outside of the 'macro factors' regarding technology, generally there is another layer of concern and inertia around connectivity through technology. As an industry, construction has never been very good at sharing information. It has, with a complicated and elongated supply chain, tended towards separation and compartmentalisation of different pieces of the commercial jigsaw. This has been seen as leading to greater control, at a main contractor level, and the creation of opportunities to manage the project in a way to minimise loss and maximise opportunity. Of course, every level in the supply chain needs control and profit to be delivered. You therefore end up, as the counterbalance to control, with cost layering which is, of its very nature, inefficient. Some common reasons for the compartmentalisation and its impact on the optimisation of ConTech for commercial gain are set out below:

- I need to pass risk and responsibility to those undertaking the work. ConTech requires a too open and too balanced structure for this.

- I don't want to share a problem until I have had time to think of a solution. ConTech shares too much information too rapidly to allow this.
- I need to apply different terms to my supply chain that I don't want to share. ConTech only really works where everything is back to back so does not enable or support this.
- My suppliers are not sophisticated enough to take part in this type of technology. If only I am using the ConTech I won't get a benefit and the investment is wasted.

There is an underlying fear and suspicion of handing over control, particularly commercial control, to systems that are not fully understood. I have labelled this the 'Terminator' effect.[42] Basically it is shorthand for the perceived dystopian consequences of handing control of our lives and processes over to artificial intelligence, robots, and ConTech only to find that the machines realise the primary cause of inefficiency is human involvement and set out to eliminate it.

Joining up not only ConTech solutions, one to another, but also integrating them into existing commercial process is a challenge. There is, however, a substantial reward. Taking the categories of ConTech described in Section 20.7, achieving meaningful use and integration of each category could lead to a scenario such as the following project.

A business park is being developed out of town. The site was surveyed by drone at tender stage and the information made available, including laser levelled 3D ground profiling and general indications of the subsurface from ground penetrating radar that identified certain areas of variability in the ground formation. The contractor was able to define the risk and price it.

The tender documents were all digital and integrated, being fully searchable and electronically cross referenced. The tenderers were able to run that tender information through their intelligent analytics systems that compared the information and risks to the historic information across the business, compared the risk profile to other current tenders and assessed the total level of risk across the business, generating a 'risk pot' of suggested value to be included in any tender to balance business and market risk. The tender profile was compared across market norms. In addition to the risk, the measure was verified automatically, comparing drawings and quantities along with other descriptive information, and cross checked with the drone imagery of the site. Baseline costs for labour, plant, and materials were established through data searches of current and recent jobs, and compared to the general market 'temperature'. An initial programme for the works, resource and cost loaded, was prepared automatically and a risk profile linked to that programme was also generated. The commercial Bid team were able to take informed and specific views, not just on the risk-return from the individual project, but also on how that fitted with the risk–return profile across the business as a whole, allowing them to give a focused view on pricing beyond cost.

At contract award a digital design was completed and modelled into a 3D environment utilising the drone data. It was linked to the programme so that construction of any part of the project could be viewed anywhere along its timeline. As the commercial team had generated a risk profile tied to the programme they were able to look specifically at the state of construction at the risk peaks and review proposals from the AI system as to how those peaks could be levelled out to provide a more consistent risk profile. The commercial team were able to utilise resource located throughout their network

---

42 Depending on age, the reader may feel more comfortable with 'i-Robot' or any number of other pop-culture references.

as all relevant information was online. Contracts with the whole supply network were prepared automatically and distributed electronically.

After the site closed each night a survey drone would remap the site and calculate progress automatically. Once buildings started to be closed an autonomous vehicle carried out the same function internally. The progress recorded was analysed across the productivity expected against the programme and any anomalies highlighted for review. Progress was also cross checked with information generated from the tags in all labour PPE, plant and equipment as well as the tags on incorporated, on-site and for-delivery materials. Where progress was ahead or behind this information was automatically sent to the materials suppliers so that the right materials were delivered each day and yard space was minimised. The progress data was then analysed against the labour, plant, and materials records to generate an earned value analysis (EVA),[43] which was cross checked against the programme and used to update the payment schedule. The commercial team, not based on site, were able to review and consider the flagged issues generated overnight from the analysed data and the proposed solutions were considered and adopted as appropriate for further optimisation. All the progress data was made available across the whole team on a real time basis.

During the project one of the areas of variability identified by the drone at tender stage proved too soft for the intended trench foundation design and a change to a piled design was required. As soon as the borehole confirming the soft ground was complete, the data and analysis of the soil, provided electronically on site, was shared with the team. The information was fed into the BIM model and a proposed design alternative, using standard engineering principles and the other design information in the model, was generated the same day ready for verification. The contractor's AI automatically produced an early warning and, using market norms, estimated the time and cost required and the likely impact on the programme and cost. The commercial and design teams were able to review the standardised proposals within 24 hours and accept the proposals with minor changes. The programme was then automatically updated to minimise the time and cost impact of the change, and consequent changes were issued to all members of the supply chain impacted by the change. A contract was issued to a new piling contractor with full design specifications.

As progress, cost and EVA information was being uploaded and verified every day, the payment cycle was weekly and the systems automatically generated the required applications and notices. The employer's systems verified these applications from the same base data and automatically confirmed payment could be released, allowing cash to flow through the supply chain within two days of application. Where any matters were flagged as not agreed (for example the cost and value generated through the change to piled foundations where allocation of responsibility had not yet been agreed) an appropriate notice was produced automatically identifying the exact cost being withheld, giving reasons. The commercial team were able to focus their efforts on any areas of disagreement based on a common data set.

After completion of the steel frame to one unit, the development productivity of the frame erection team for the next unit was recorded as dropping 10% below the expected level for three days in a row. The system flagged this to the commercial team, together

---

43 EVA: an analysis of costs incurred against value generated in terms of amount of construction work completed.

with an interpretation of why that might be the case based on the measurement data for amount of frame erected and other site data such as time, location, and activity levels recorded from the PPE tags of the relevant operatives. On this occasion the system identified that there had been three days of heavy rain and the drone recordings showed standing water on site where the steel erectors should have been working. The system reviewed the weather forecasts, which were for more heavy rain and then provided a cost-risk analysis of accepting the lower productivity, adjusting the programme to work in a different area or dewatering that part of the site, including availability of equipment (internally and in the market) to enable that. The commercial team were able to review the collated data and proposals, engage that day with the subcontractor and agree a way forward sharing cost and risk.

At the end of the project there was, in fact, very little discussion required to finalise the account as matters had been closed progressively throughout the supply chain. That freed the commercial team to look at what worked and did not work, and provide analysis on improvements both to the systems used and what further ConTech support could have been deployed. The entire project data was analysed by the AI and trend analysis provided across the project and business comparing initial forecasts with out-turn. The entire data set was then archived as a data source for future project analysis.

Every decision and proposal made by the AI system, together with the data that was fed into it, was logged, whether or not proposals were implemented. This allowed a full business risk review to take place on whether an advantage was obtained through use of the system. Information on the project was compiled into a report suitable for Board directors and the key statistics, as they had been during the project, compiled to provide the necessary data to both insurers and auditors.

The key overall benefits to the commercial team of joining the ConTech together were that a smaller, more agile team was able to take earlier and more informed decisions on the project while understanding the impact of those decisions on total business risk. It allowed optimal utilisation of commercial resource by not having specific people locked to particular projects but enabled a wider group to work on a portfolio of projects from any locations in the world. That enabled rapid deployment of specialist resource for different stages of the project. The focus of the commercial team was on implementation and testing of analysis rather than data searching and compilation of reports. The skill sets for the team were different, as were the demands on them.

20.9 Project Control and Risk Management – The Future

Having looked at all the different types of ConTech available, and their individual and collective use and benefit to commercial management, there is still one thing missing; one piece of technology that has not been discussed in detail but is a hot topic for any technology debate: Blockchain.

As a piece of technology a Blockchain has very little, at face value, to do with a construction project or the commercial management and delivery of those projects. At its simplest a blockchain is a connected sequence of verified activities that is very difficult to alter once in place. It is also a decentralised system, meaning that there is not one single data source but multiple co-existing identical copies of equal validity available to anyone taking part in the Blockchain. Converting that into a very simple construction context,[44] a Blockchain is like a works programme with links from one activity to the next; however, in a Blockchain every link must be F–S.[45] As each activity is completed and verified as complete, so a new block is created in the Blockchain. In this way a single activity in the programme could be broken up into multiple blocks in the Blockchain.[46]

The importance of the decentralised storage is that:

- Any attempt to alter the data being recorded would have to simultaneously change multiple copies of the record and all subsequent blocks, which is difficult and unlikely, thus high levels of security and trust in the information held in the Blockchain is generated.
- It provides for transparency as any member of the project involved in the Blockchain can review it at any time and be certain they are looking at the same data set as everyone else.

It is perhaps interesting that the key characteristics of a Blockchain, and what makes it so revolutionary and interesting in a financial service context, are reasons why it is likely to be dismissed by the majority of those in the construction industry. As noted earlier, there is a huge reluctance within the construction industry to share information. Without that acceptance of sharing, a decentralised system such as a Blockchain cannot work. The question then is whether the advantages of a Blockchain (or sharing information generally) are great enough to outweigh the industry inertia and reluctance to share.

What are the advantages, or what could be achieved by putting a construction project into a Blockchain, what would it look like and how would you do it? As always, the starting point, particularly from a commercial perspective, must be 'what is the problem we are trying to solve?'. Let us take three problem areas in particular and examine what Blockchain could offer.

The first and most obvious issue a Blockchain could impact is payment. Once you have a system where work is verified and confirmed, based on shared data, payment can be released very quickly. The second is auditability of the project decision making. Where everyone can see what has happened in close to real time[47] you take a large step closer to a common data set and a common understanding of the facts of the project, and what happened and is happening. Third, a Blockchain can reduce the need for layering of commercial resource through the supply chain. As there is one common data set accessible to everyone, the need for separate commercial processes, and the inevitable cost and opportunity for disagreement that flows from that, can be dramatically reduced.

These are arguably three of the biggest issues facing commercial management in construction at the moment, all of which can be positively influenced by a decentralised management system such as Blockchain.

How then would a Blockchain actually work for a construction project? The theory is relatively easy: a set of criteria would be established for completion of an item of

---

44 There are many further levels of development and deepening of the use of Blockchain as a principle that shall be explored.

45 A strict Finish to Start logic link.

46 The level of granularity of the Blockchain can be set by the participants of the Blockchain to achieve the right balance of detail for the project.

47 You could reasonably expect a Blockchain to be updated every few hours to understand what is happening on site.

work. Those criteria would reflect, in a binary (yes–no) fashion the requirement of the contract. On demonstration of the criteria having been met a block can be closed. Once a block is closed, and closure requires verification by the participants in the Blockchain, payment and other consequences can flow automatically. The practical difficulty is in setting the criteria, as that will need to identify all the questions that need to be asked in order to confirm that the work has been done in such a way that it is complete and payment can be made. Setting the criteria requires some thought to optimise the process but can certainly be done. Taking as an example a brick wall, there will need to be data around time, date, and location of the work, there may be a minimum work content requirement to size a block (1000 bricks or every lift depending on the project) and financial confirmation of the rate per unit. The more challenging requirement is demonstration that the bricks have been laid correctly with the correct materials, but that could be material tracking and a confirmation photograph or video that can be reviewed. Once that is in place and the block is closed anyone else can see that block is in place and will understand what it means to their own work package.

The opportunities around adoption of a Blockchain as a method of framing a project and its use of ConTech are significant, and as yet relatively untapped, but it is a good mechanism for adding the controls to a ConTech driven system.

20.10   Conclusion

There are enormous opportunities presented through ConTech for improving commercial management but in order to realise those opportunities you need to go beyond the technology itself and re-evaluate the business issues and inefficiencies you need to deal with. ConTech needs to be challenged for its commercial value and delivery; many systems promise a lot but deliver little in terms of true commercial value. Understanding what is out there and how it could be adapted or utilised to serve a commercial purpose is key as the market is moving fast.

21

Intelligent Document Processes to Capture Data and Manage Risk and Compliance

Graham Thomson, CEO at Affinitext Inc.

21.1 Introduction

A clear trend in developed economies is for organisations in both the public and private sectors to enter into increasingly large and complex contracts for the supply of goods and services.

This increased complexity is not only of great concern to the success of projects, but is also of existential concern to the companies involved. As PwC have said: *'You cannot claim to be in control of your business if you are not in control of the contracts it depends on'.*[1]

Putting a real-life perspective on this complexity, one of the top 10 construction projects of the twentieth century was the Hoover Dam in Nevada. The contract for the Hoover Dam was 172 pages long (100 pages of text and 72 drawings). There were eight defined terms in the contract, and less than 200 contractual obligations.

By the late twentieth century, and into the twenty-first century, deal-closers developed their 'art' to the point where a Public Private Partnership (PPP) contract suite for building a school, hospital, road, etc. was 13 500 pages long, with 5000 defined terms (referenced 100 000 times through the contracts), 16 000 clause–clause or other cross references and over 4500 contractual obligations or entitlements. This complexity is what the deal-closers and their legal advisers pass over to the contract management team to manage for up to 50+ years. After his deal-close euphoria subsided, the project director on the largest PPP project in the UK referred to the sheer complexity of the contracts for the project as being an 'unintended tidal wave of risk', which required prompt management to bring it under control.

So how does management resolve this complexity problem? In answering this it must be recognised that, while human resources are core to the success of complex contracts, they are one limb of the three-limbed formula for successful projects:

(People▼) + (Best Tools and Processes▲) = Successful Projects

A simple truth is that most projects cannot have the very best skilled or experienced people in every role, either because they are not available or the project is unable to

---

1 PwC. 2003 *Contract Management: Control Value and Minimise Risks*. PwC Report. p. 18.

afford them (a reality that has become starker with the ever growing pressures on organisations to 'do more with less'), and the problem is exacerbated by the fact that people move on. When they do, their knowledge and experience is largely lost on the project they were engaged on, let alone captured and shared for the benefit of other projects within an organisation's portfolio. Given the inevitability of these people issues, it is common sense to invest in the best tools and processes to ensure a detailed understanding of the contracts, as well as to capture and share knowledge, and manage risk and compliance.

As in all walks of life, technology is rapidly evolving to meet the people and complexity challenges referred to above. In relation to the engineering, design, construction, and asset management aspects of projects, building information modelling (BIM) software is now seen as an essential part of the future. In the UK the *Government Construction Strategy 2016–2020*, re-states the Government's 2011 commitment to require collaborative 3D BIM Level 2 (with all project and asset information, documentation, and data being electronic) on all its projects by 2016 and to develop, in conjunction with industry, the next generation of digital standards to enable 3D BIM Level 3 adoption (with full collaboration between all disciplines by means of using a single, shared project model) under the remit of the *Digital Built Britain* strategy. 3D BIM will evolve into nD BIM (augmenting the three primary spatial dimensions (width, height, and depth) with time as the fourth dimension (4D), cost as the fifth (5D) and facilities management (FM) as the sixth (6D).

In the same way as BIM was developed for drawings, equipment, and space, a new twenty-first century intelligent document format (IDF) has been developed for textual documents such as complex contracts, guidelines, regulations, reports, etc., which underpin any project. IDF is a very powerful use of HTML and associated technologies.

A little bit more history. Again, going back to the twentieth century, MS Word was released 35 years ago by Microsoft for the purpose of authoring and editing documents. It is fit for that purpose. Still in the twentieth century, 25 years ago, PDF was released by Adobe for the purpose of sealing and printing documents. It is fit for that purpose. Both these formats are 2D. Both were developed pre-Web. Neither of these 2D formats is fit for the purpose of understanding complex contracts and associated knowledge and business processes.

While the construction industry is notoriously behind other industries in implementing new technology, a simple move from 25 year old PDF technology to current day IDF technology immediately transforms the understanding and management of complex contracts with:

- On-screen navigation with pop-up definitions and clause–clause linking within and between documents.
- Fully conformed, up-do-date versions of the contracts.
- Uniquely powerful search ability.
- The ability, within seconds, to extract all contractual obligations and entitlements from very complex projects and publish them into an intelligent spreadsheet, which enables users to decide which they want formally to manage, by whom and when.
- The ability, within seconds, to use that completed spreadsheet to create tasks at each relevant clause, with 'just-in-time' reminders being sent directly from the clause, and auditable compliance available both directly at the clause and in a real-time database.
- Knowledge capture and sharing directly at the relevant clauses.
- The ability to capture, share, and annotate legal analysis throughout the project documents.

IDF is currently 4D, but intelligent document processes are evolving to 5D (cost) and 6D (predictive).

21.2   The Dimensions of IDF

The Dimensions of IDF are:

IDF Dimension	Description
3D	The 3D component of IDF is the depth of document reading and understanding. IDF provides web-based 3D navigation of documents on PCs, tablets, and mobiles with: • All defined terms being pulled up as pop-ups at the click of a mouse wherever they are referred to, no matter whether they come from the agreement in question, or from another document. • 100% pinpoint hyperlinking of all clause-to-clause references or other linkages within and between documents. • Interactive, full-text, relational database searching within and across projects. • Contract management guidance from the contract management manual linked directly from the relevant clauses. • Seamless integration with business/social media applications.
4D	The 4D component of IDF is time. Documents and associated knowledge and business process evolve and change over time. IDF is a dynamic format. IDF interactive clauses push information, knowledge, and tasks to relevant users as and when they need it. The clauses become an active part of project delivery, making the complex both manageable and usable: • The agreements are always fully up to date, with all amendments incorporated, but with an icon linking to the amendment history of any paragraph. Timeline sliders will allow you to slide your contract back through its history. • Any relevant knowledge is pushed to a user as they view a clause (whether it be minutes of a meeting as to how the parties agreed the clause is to operate in practice, a legal opinion, technical report, flow chart, etc.) • Tasks/obligations that need to be performed are attached directly at the relevant clause, with 'just-in-time' reminders being automatically sent to the relevant person direct from the clause. A full audit of the task performance, along with verification and compliance evidence, is captured behind the clause as the steps of task completion take place.
5D	The 5D component of IDF is cost. Significant investment is being made in 'smart contract' blockchain processes to improve payment times and reduce administrative costs of payments through the supply chain, which can link seamlessly to the contract.
6D	The 6D component of IDF is predictive technology. There are exciting opportunities arising from this; some quicker to achieve than others. Examples include: • Hot spot detection and heat mapping, with alerts sent for the contractual issues that are looked at currently and over time across portfolios of projects, giving early warning as to potential issues and or areas in which training may assist. • Extracting information from data sets such as expected or actual rainfalls and plugging into project management software (e.g. Prima Vera), and connecting with contractual notice clauses such as delays for inclement weather. • Detecting trends or patterns in changing risk allocation in project, construction or finance documents on new projects. • Implementing blockchain technology

While the benefits of IDF are self-evident for any high-value documents, a further value of IDF for projects arises in the context of the new ISO 44001: Collaborative Business Relationships. ISO 44001 certification is now rapidly being sought or required on major projects in the UK. It utilises an eight-phase model, from project concept through to disengagement, covering the 'people', 'tools', and 'processes' elements of the formula for successful projects referred to earlier, see Figure 21.1.

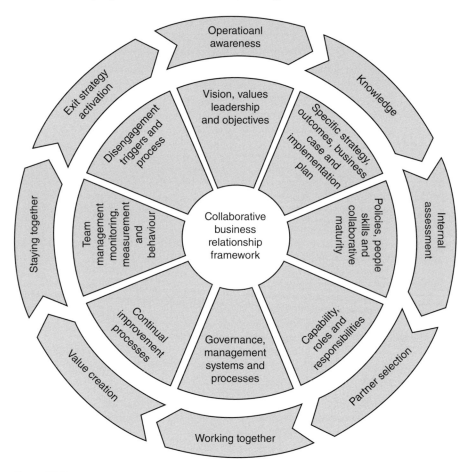

Figure 21.1

Certification requires the project to focus on, amongst other things, relationship management, effective sharing of knowledge, innovation and management systems, and processes.

Before intelligent document processes, parties to a contract generally managed them in silos, each party having its own version of the contract, hopefully with all its amendments, its own spreadsheet or database of contractual obligations, its own contract management manual (at least in theory, if not in practice!), etc.

ISO 44001 has increased industry recognition that the contract is the glue that is binding the parties together to deliver agreed objectives for agreed rewards, and that a silo mentality to the management of the contract is inefficient, risky, and harmful to establishing the best relationships.

With IDF and a focus on the people collaborating effectively in projects, best practice is now as follows:

- All parties have online access to the same fully up to date version of the contract in IDF.
- The parties agree a common contractual matrix of obligations: who needs to do what, by when. 'Just-in-time' reminders of these obligations are sent to the responsible user direct from the relevant clause. Once the obligation is completed by the relevant person, an automated email goes to the verifier from the other party for confirmation that the obligation is satisfactorily performed. The compliance information now sits behind the clause, along with a complete audit trail of its completion.
- When the parties agree how a clause is to be interpreted in practice, that agreement is captured and shared at the clause for the benefit of new staff, advisers, etc. The same applies for knowledge within an organisation; turning personal knowledge into corporate knowledge.
- The ISO 44001 standard, a compliance map, and the corporate relationship management system and joint relationship management plans are converted to IDF and the standard is implemented dynamically to embed collaborative working into the DNA of the project, in the same way as health and safety is.
- There is a single, common contract management manual or guide that the parties have agreed between them for the management of the project; again, embedded at each relevant clause of the contract for quick reference. In the PF2 model contracts in the UK, the Government mandates that the contractors provide this manual.
- Using social media tools, team discussions can be integrated seamlessly with the relevant technical or contract clauses.
- For organisations with portfolios of projects, the collation of a library containing all projects instantly provides an extremely powerful knowledge hub for the organisation.
- For Government contracts, the public has access to an online redacted version of the contract in IDF. This is 5* transparency in Open Government.

Using IDF is a quick, easy, and extremely powerful solution for achieving each of the above, and powerfully positions organisations for the introduction and rapid growth of further new technologies such as AI, the IoT, and blockchain, which are all currently being heavily invested in (US$2+ billion is expected to be invested in blockchain technology alone in 2018).

To conclude with my own mantra for excellence in project and contract management, which necessitates ensuring that competent people and the best tools and processes are seamlessly integrated at the outset of a project:

Do it once: do it right!

22

Organisational Information Requirements for Successful BIM Implementation

Dr Noha Saleeb

Associate Professor in Creative Technologies and Construction, Design Engineering and Mathematics Department, School of Science and Technology, Middlesex University, UK

Abstract

Success of the business and its profitability is the ultimate goal for any organisation in the construction industry, whether that is achieved through return on investment (ROI), reputation, repeat business or client satisfaction. To accomplish that, an organisation must have a well-defined business plan and objectives to implement. This chapter discusses how a successful business plan can be laid out through developing Organisational Information Requirements (OIR) in tandem with usage of Building Information Modelling (BIM) management. This will aid decision makers at corporate level to devise a business management strategy that can ensure success of running their developed assets to highest efficiency and ensure best ROI from them. This chapter hence provides a step-by-step process of how to create the Organisational Information Requirements documentation (OIR).

22.1 Introduction

Building Information Modelling (BIM) is a concept that has been adopted by the construction industry in an endeavour to ultimately migrate to a digital economy and hence achieve best efficiency, which would eliminate waste inherent in the industry projects and deliver best return on investment. It has been given many definitions, e.g. 'a process for creating and managing information on a construction project across the project life cycle' (NBS 2016). Furthermore, this chapter adopts a more holistic vision of BIM as an overarching umbrella that encompasses within it usage of more efficient processes, workflows, roles, standards, policies, and digitalisation to maximise benefit and profitability of construction or infrastructure projects via information management enabled by technology (Figure 22.1). This can include different types of management such as design, construction, project, supply chain, time, cost, risk, and quality management. Reflecting on all these components, it becomes logical that for a construction organisation to successfully manage all these on all its built assets, these categories should be embedded within the organisation-wide long term strategic business plan in order to define what needs to be managed, the business objectives to be achieved, and key

Global Construction Success, First Edition. Charles O'Neil.
© 2019 John Wiley & Sons Ltd. Published 2019 by John Wiley & Sons Ltd.

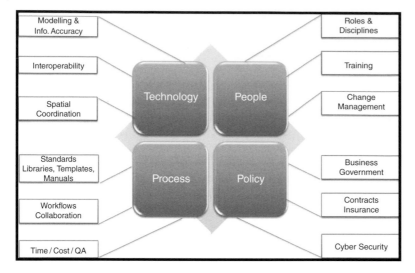

Figure 22.1 Components of building information modelling.

performance indicators (KPIs) to measure them, as opposed to managing each asset ad hoc individually with no overall vision for successful outputs and outcomes.

This merge between organisational business goals/objectives and BIM necessitates the creation of Organisational Information Requirements (OIRs), as will be explained subsequently.

22.2 Leveraging Organisational Information Requirements for Business Success

22.2.1 Business Success

It is vital to first determine the importance of thoroughly defined OIRs for the success of organisations, which is detailed below. This depends on many factors, e.g. improved business planning decreases the risks accompanying any business activity. Lack of new technology and equipment are also hindrances in development. A recent study by Jasra et al. (2011) developed and confirmed a framework for the relationship between small and medium-size enterprise (SME) business success and its determinants. The study determined that financial resources, technological resources, government support, marketing strategies, and entrepreneurial skills have positive and substantial effects on business success (Figure 22.2).

Another example arises where Teo and Ang (1998) demonstrate that coordinating information system (IS) plans with business plans is one of the main difficulties in IS planning. Three critical success factors determined to achieve this are:

- ISs management is knowledgeable about business
- Business goals and objectives are made known to IS management
- The corporate business plan is made available to IS management.

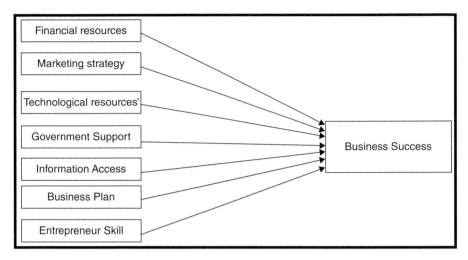

Figure 22.2 Framework for the relationship between business success and its determinants (Jasra et al. 2011).

Furthermore, changing an organisation's focus to business objectives and the whole of life risks/rewards are of vital significance to reducing overall risk. Evaluation of risks must be founded not only on producing projects on time and within budget but also on creating and managing a long term business plan that can deliver the business objectives of the organisational stakeholders while fulfilling community expectations (Jaafari 2001).

The previous cases highlighted the importance of identifying organisational needs/requirements and business objectives for overall business success. Section 22.2.2 clarifies how this can be identified within the OIR documentation created as part of the BIM strategy for creating built assets and projects.

22.2.2 Organisational Information Requirements (OIRs)

OIRs are defined in PAS 1192-3:2014 – Specification for information management for the operational phase of assets using BIM (BSI 2013). OIRs explain the information needed by an organisation for asset management systems and other organisational functions/needs. Hence they are organisational-level information requirements as opposed to asset-level or project-level information requirements, which are created based on the OIR as per Figure 22.3. Furthermore, OIRs help inform the plain language questions asked by the client of the supply chain in the employer's information requirements (EIRs) document to achieve the project and create its BIM implementation documentation such as the BIM execution plan (BEP). The resulting outcome of this should be to achieve more efficient projects with least risk possible, successful business objectives, and return on investment.

PAS 1192-3:2014 Annex 2 (BSI 2013) provides generic guideline activities that can aid in the creation of the OIRs; however, these require further information on how to achieve them or what to ask for as information or deliverables in the OIRs to actually accomplish that. Table 22.1 created by the author elucidates the minimal clarification

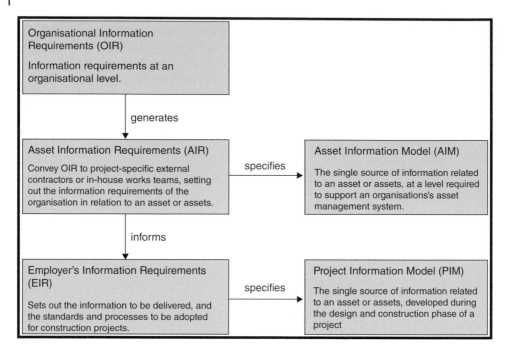

Figure 22.3 Relationship between the elements of information management (BSI 2013).

categories required to include the generic guideline activities suggested in PAS1192-3 in the OIR documentation to be able to produce the information required by the organisation.

Furthermore it can be noted that the suggested activities have no aspirational objectives or goals (except 'optimising asset strategy', which is generic), but merely 'identify'/'measure'/'assess'/'determine'/'obtain' information about certain tasks with no enhanced business outcome required as a result of them. These in their own right might be classified as 'OIRs' to produce, but they do not satisfy any particular business plan for the organisation, hence cannot be used to gauge the success of that business plan.

It is therefore necessary to know how to create an OIR with specific defined goals to best serve the purpose of achieving business objectives using BIM methodologies, which is explained step-by-step in Section 22.3.

22.3 Developing OIRs Using BIM

As demonstrated previously, it is imperative to embed the goals and objectives of the organisational strategic business plan within the OIRs and to request that information about these is delivered as a method to ensure achieving the goals. This will enable having benchmarks to track performance, reduce risks and make interim corrections during the implementation of different projects advocated as vital by Sadgrove (2016). Furthermore this will allow analysis of revenues, costs, and projected profits, and hence the organisation's strengths, weaknesses, threats, and opportunities.

Table 22.1 Organisation activities as per PAS1192-3 versus required information categories for implementation in the OIR.

Organisation activities	Information clarification required for implementation in OIR
Optimizing the asset management strategy	• Benchmark for optimisation • Categories of asset management
Assessing financial benefits of planned improvement activities	• Benchmark for benefits • Categories of improvement
Modelling the asset to support operational decision making	• Modelling deliverables • Operational categories
Determining the operational and financial impact of asset unavailability or failure	• Categories of failure and unavailability • Benchmarks of failure • Timelines of failure and unavailability
Making life cycle cost comparisons of alternative capital investments	• Other variables required, e.g. procurement type, investment type, horizontal versus vertical integration • Supply chain involvement type
Identifying expiry of warranty periods	• Benchmarks and impact categories – too many components at organisational level • What business objectives this affects
Determining the end of an asset's economic life, e.g. asset expenditure exceeds the associated income	• What actions to perform/objective to aspire to as a result • Total expenditure/income or categories' expenditure/income – which are critical?
Undertaking financial analysis of planned income and expenditure	
Determining the cost of specific activities (activity-based costing), e.g. total cost of maintaining a specific asset system	• Benchmarks to gauge costs against • What duration of time or asset phase
Obtaining/calculating asset replacement values	• Needs to be per component, which should not be detailed at organisation level
Calculating financial and resource impact of deviating from plans on change in asset availability or performance, e.g. financial impact of deferring the maintenance of a specific generator by six months	• Needs to be per component, which should not be detailed at organisation level • Categories of plans/at what phase – what is critical • Benchmarks/categories of functional performance
Assessing its overall financial performance	• Benchmarks/categories of financial performance
Undertaking the ongoing identification, assessment and control of asset related risks	• Categories of risk • Benchmarks for controlling risks

Figure 22.4 displays the author's vision of the seven essential criteria that require delivery of information at organisation strategic level (i.e. OIRs). Figures 22.5 and 22.6 present subcategories of these information criteria that an organisation can create within an OIR, with Figure 22.7 providing an example of a template that can be utilised to write these information requirements.

Two of the seven criteria 'Asset Management Optimisation' and 'Spares, Maintenance and Purchasing' align with the criteria in PAS 1192-3:2014, which specify information

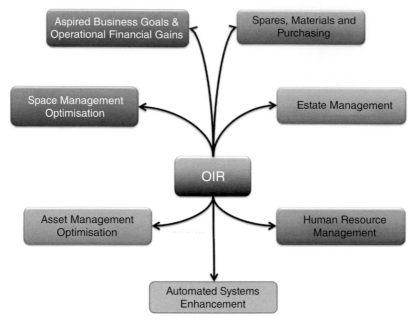

Figure 22.4 Categories of Organisational Information Requirements.

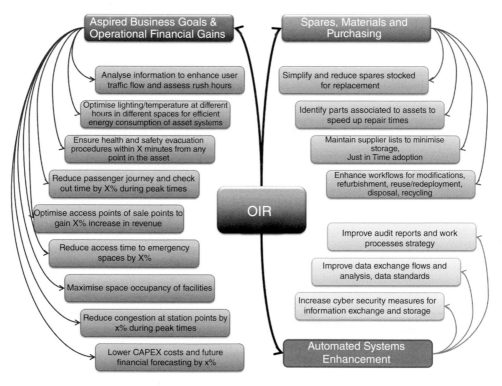

Figure 22.5 Subcategories of organisational information requirements (1).

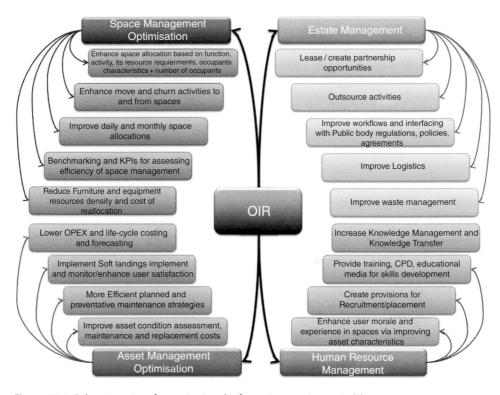

Figure 22.6 Sub-categories of organisational information requirements (2).

Organisation / Project ID	Resolution Owner/ Manager	Risk owner	Financial owner	Plain Language Question	Acceptance criteria	Supporting documen-tation	Start date	Completion date
Reducing access time emergency points by 30%	(If applicable)	(required to resolve factor or find solution for it)	(of risks associat-ed with factor)	(involved with factor if applicable)	(PLQ to define / outcome for that factor)	(to assess performance of factor) • Standards • Methods • Procedures • Tolerances • LOD/LOI • KPIs	(related to factor resolution or information)	(if applicable-or project phase)	(if applicable-or project phase)
			+ Project type + Level of Objective completion (or asset completion?)						

Figure 22.7 Example template for OIRs.

needed by an organisation for asset management systems. However, the other five criteria feed into achieving and enhancing the success of the business plan:

- 'Aspired Business Goals and Operational Financial gains' – tackles information that can affect the design, construction, and operations decisions of future assets to achieve business goals.
- 'Automated Systems Enhancement' – deals with information related to improving smart and automated systems and procedures within future assets.
- 'Space Management Optimisation' – handles efficient occupancy and habitation of spaces within an asset by the users.
- 'Estate Management' – tackles financial partnership opportunities.
- 'Human Resource Management' – deals with wellbeing, development, and lifestyle of the end users.

22.3.1   Aspired Business Goals and Operational Financial Gains

As per Figure 22.5, an organisation can seek to enhance certain functionalities in their services provided in their built assets, which in turn would deliver better return on investment and profitability. An example of this is a developer who creates a series of similar type assets, e.g. supermarkets, and desires to reduce the end user's time for purchasing or checking out by 20%; airport terminals where passenger journey and check-in times need to be reduced by 30% to reduce traffic at peak times and enhance the experience; hospitals where time to surgical wards or patient waiting times in A&E must be reduced by 50% to save patients; tube stations or office buildings or schools where the evacuation time in case of emergencies at peak times needs to be reduced to three minutes. The aforementioned examples relate to achieving time-based business goals. Other business goals can be cost based, e.g. increase revenue from sales or food/beverages or hiring spaces by 10%; reduce cost of paying for energy consumption, heating and lighting in a building by 30%; reduce CAPEX costs by 20% by automated construction methods. All of these business goals, whether cost or time based, would have a direct impact on the design of the spaces within the assets involved and their construction methods, to achieve the goals, hence the relationship with creating AIRs and EIRs, as OIRs will dictate how the project is run and created. Ensuring this relationship will reduce the risk of failure of the business plan/venture that the organisation has in place. It must be noted that usage of metrics (e.g. 20%, etc.) provides a benchmark that the goal can be measured against (KPI), and helps designers and constructors to achieve a certain standard of quality in their work. The early coordination between these disciplines using BIM methodologies of efficient digitised information exchange, modelling, synchronous working, automated scheduling, off-site manufacturing, visualisation, and simulation of options and alternatives will enhance this process and ensure achieving it to the highest quality possible with least risk.

22.3.2   Spares, Maintenance, and Purchasing

The subcategories of this criterion as per Figure 22.5 consider enhancing the logistics of performing maintenance by optimising stored spares and hence reducing costs. This could also be achieved by efficient analysis of suppliers and their benefits or provision of just-in-time supplies to reduce warehouse storage and risk of damage from faulty or long storage. Sharing resources between projects could also be a solution. All these cost and time related goals require actions to be taken during the construction phase and hence informed decision making for this phase. Again usage of BIM project management methodologies of visualised scheduling, automated cost estimation and analysis, and performing time and cost clash detection (not just design) will enhance achieving these goals and reduce the risk of organisational business plan failure.

22.3.3 Automated System Enhancement

The subcategories in this criterion, shown in Figure 22.5, require transforming construction projects to a highly digitised environment, which is enabled using BIM methodologies. To prevent corruption, loss and piracy of data, better cyber security methods should be implemented organisation wide. Improvement of data analysis, audits, exchanges, ensuring standardisation of components and information are all goals that should be aspired to, as they render processes and workflows more efficient and less prone to risk of error and hence losses and failure. These could include usage and management of smart telemetry in smart assets, big data analysis, Internet of Things, robotics, artificial intelligence, and other industry 4.0 technologies. It can be noted here that, unlike Section 22.3.1 above, no benchmark metrics are being used here due to the fact that 100% improvement in these categories (cyber security, data analysis, exchange, quality audits, etc.) are what is aspired for here.

22.3.4 Asset Management Optimisation

As per Figure 22.6, this includes enhancement of planned/preventative and reactive maintenance strategies, forecasting problems with the different asset systems, implementing the Government Soft Landings strategies to ensure better quality post-occupancy evaluation, hence reducing overall OPEX (operational) costs. Again this can be more efficient using BIM methodologies, which use effective data analysis of the performance of the different systems to monitor and maintain them, virtual and augmented reality, e.g. for computational fluid dynamics (CFD) simulations, etc.

22.3.5 Space Management Optimisation

Operations management of an asset does not just involve maintenance of its systems; a very profitable aspect of an asset is efficient management of its spaces, e.g. distribution of lecture, class, and meeting spaces in office and educational facilities. These can be optimised not just according to numbers of occupants but also according to the function and activities of the gathering, the resources required for it, and even the characteristics of the occupants. Furthermore control of smooth movement (churn) between spaces is vital for both people and resources to transport between spaces. Again different discipline collaboration on analysis of data and visual analytics using BIM methodologies can aid this process.

22.3.6   Estate Management

This criterion looks at a broader picture, as per Figure 22.6, in terms of attracting funding opportunities, which can be public, private, using horizontal or vertical integration of organisations and their supply chains, and the benefits/drawbacks of each. These aspects can affect how a project is implemented and an asset built, hence requiring thinking of suitable procurement types, e.g. more collaborative integrated project delivery or insurance (IPD/IPI) routes, which will ultimately change how a project is managed and the BIM methods/tools/procedures/roles required to facilitate that. Other factors included in this criterion are enhancing logistics and waste management processes.

22.3.7   Human Resource Management

This final criterion is probably the most unusual to include in a business plan and require information for. However, considering that the ultimate goal of any business is repeat business, which can only be achieved by customer loyalty and satisfaction, it becomes apparent that having goals related to end user wellbeing and satisfaction is crucial. Factors for the organisation employees include providing training, continuous professional development (CPD), placement opportunities, and knowledge transfer. Factors for the asset users include enhancing their experience within the spaces to achieve higher quality experience and improved morale, which can be achieved by improving the design of the asset. This engagement between supplier and end user is part of the BIM collaboration process, and helps reduce the risk of failed projects.

22.3.8   OIR Template

Upon identifying all the criteria and subcategories of information that can be included in an OIR, it is imperative to find an understandable way to translate this into document form. Figure 22.7 gives an example of a chart template that can be used to populate OIRs. It starts with identifying the subcategory at hand in the first field, followed by standard information of organisation name/ID/owner. The next field could show who, or which roles, is responsible for handling or decision making or providing the information on a particular issue, and who is financially responsible for it (could be the same or different personnel). It could then be important to write the plain language question(s) that the organisation should ask to ensure achieving an issue, followed by acceptance criteria for it. These can include KPIs, standards, benchmarks, methods, procedures and workflows, and even tolerances and levels of detail and information to be provided. It is also important to provide links for any necessary external documentation related to an issue. The start date and completion date are optional fields in case this OIR is specific to a particular project only and not a long term business plan. Extra fields could be added to show the type of project, specific characteristics related to it, etc. One beneficial field could be to indicate the current level of objective completion of a particular factor, e.g. current evacuation time. This could be useful to show what needs to be done to reach the new objective metric, e.g. three minutes, 20% enhancement, etc.

Ultimately, the way an OIR document is expressed can be in any format that the organisation sees as being most viable and efficient to express its needs.

22.4   Conclusion

This chapter discussed the importance of aligning an organisation's strategic business plan with its information requirements to achieve its business goals and objectives successfully. Furthermore, a step-by-step guide for how to create information criteria and subcriteria was recommended to populate an OIR chart and ask for relevant information using the methodologies of BIM. This enables executive decision makers to create a strategy that can inform the creation of Asset and Employer Information Requirements documentation (AIRs and EIRs) for individual projects, and influence design, construction and operation decisions during their respective phases to achieve the overall business goals successfully.

References

BSI (2013). *PAS 1192:2 2013 - Specification for Information Management for the Capital/Delivery Phase of Construction Projects Using Building Information Modelling*. London: BSI.

Jaafari, A. (2001). Management of risks, uncertainties and opportunities on projects: time for a fundamental shift. *International Journal of Project Management* 19 (2): 89–101.

Jasra, J.M., Hunjra, A.I., Rehman, A.U. et al. (2011). Determinants of business success of small and medium enterprises. *International Journal of Business and Social Science* 2 (20): 274–280.

NBS. (2016). What is Building Information Modelling (BIM)? Available at: https://www .thenbs.com/knowledge/what-is-building-information-modelling-bim. (Last accessed 13 May 2018).

Sadgrove, K. (2016). *The Complete Guide to Business Risk Management*. Routledge.

Teo, T.S.H. and Ang, J.S.K. (1998). *Critical Success Factors in the Alignment of IS Plans with Business Plans*. Faculty of Business Administration, National University of Singapore.

23

Examples of Successful Projects and how they Managed Risk

23.1   Introduction

There is no better way to learn how to run a successful project than to participate in one personally. If this is not possible initially then hearing how to do it from highly successful senior managers is the next best way.

In this chapter of the book we are privileged to include the following two papers by Ian Williams, a recognised industry leader of major projects, describing how success was achieved on two of his high profile projects:

- *People, People, People*. London 2012 Olympic and Paralympic Games
- *Managing Risk*. Tunnels for Heathrow's Terminal 5.

These are followed by a really interesting paper by Stephen Warburton:

- *Cyber Design Development*. Alder Hey Institute in the Park, UK.

Stephen is a highly accomplished design manager with wide international experience.

23.2   People, People, People – London 2012 Olympic and Paralympic Games

Ian Williams MSc, CEng, FICE, FCIArb

Former Head of Projects for the Government Olympic Executive, London 2012

Success for the London 2012 Olympics could be measured in many ways, such as the sell-out of the tickets; the quality of the sport; the transport operations met all demands; and the security operation ensured no incidents affected the Games. All commentators, without exception, heralded the 2012 London Olympic and Paralympic Games as a great success. Ultimately the summer of spectacular sport and how London performed as a city is what most people remember. However, behind the sport was approximately seven years of preparation covering the construction of Olympic Park and its competition and non-competition venues. This preparation was the foundation for regeneration of East London, the transport and security operations, the public services and the planning and execution of the Games. The National Audit Office (the independent UK government body scrutinising public spending) stated in their Post Games Review Report that

Global Construction Success, First Edition. Charles O'Neil.
© 2019 John Wiley & Sons Ltd. Published 2019 by John Wiley & Sons Ltd.

'The successful staging of the Olympics has been widely acknowledged'....and... *'By any reasonable measure the Games were a success and the big picture is that they delivered value for money'.*

This paper covers the factors that led to the success of the activities that were funded by a £9.3 billion public sector package, with the building and commissioning of the Olympic Park and its venues making up about 75% of the budget. These key factors included putting the right governance in place, agreeing a comprehensive and realistic budget, effective procurement measures, effective progress monitoring and risk management and above all, people with the right skills to deliver.

The Olympic Park covered about 250 ha of what would be in legacy a new urban Park, which has about 100 ha of open space and which created one of Europe's largest urban parks for 150 years. Over 200 buildings were demolished and 98% of the demolition materials were used or recycled; using five soil-washing machines about 80% (1.4 million cubic metres) of contaminated soil was cleaned and reused. The Olympic Park included several permanent and temporary venues; the 80 000 seat Olympic Stadium and warm up track; 17 000 seat Aquatics centre; the Water Polo arena; 6000 seat Velodrome; 12 000 seat Basketball arena; the Hockey arena; 7000 seat Handball arena; Paralympic Tennis centre; the International Broadcasting and Media Centres; and last but not least the 3600 bed Athletes' Village.

Whilst the construction risks in themselves were challenging, the major risk identified at the outset was the ability of government to deliver infrastructure within the allocated budget and to an immovable deadline. In 2005, on the announcement that London was the successful city, the immovable deadline of 26 June 2012 for the opening ceremony was fixed. In 2007, it was identified that the main areas of risk that needed to be addressed were the need for strong governance and delivery structures given the multiplicity of organisations and groups involved; the need for a budget to be clearly determined and managed, applying effective procurement measures; the need for effective progress monitoring and risk management arrangements; **and above all people with the right skills to deliver.**

At the outset, the government appointed a specialist recruitment consultant to source key individuals with proven major project experience, and expertise in commercial and financial aspects of major projects. A number of key characteristics were sought; a clear understanding of the balance between value and cost, and individuals with a 'can do' attitude who were resourceful, who would find solutions and were collaborative in their approach.

In construction terms, achievements on the Olympic Park were extraordinary and had placed the British construction industry at the forefront of delivering major projects. The Park was delivered to schedule and within budget, with the anticipated out-turn cost of the Park, infrastructure and the Village at £7.2 billion.

London 2012 did not escape the credit crunch. In 2008 the Athletes' Village PFI (PPP) deal collapsed and the government had no alternative option but to fund its construction. However, evidence of the attractiveness of the regeneration and the strong commercial demand for the Olympic assets in legacy mode has since been demonstrated by the deal with Delancey and Qatari Diar for the Athletes Village.

23.2.1   Governance

A single government department was made responsible for the leadership of the public sector funding package, the Department for Culture, Media and Sport through its Government Olympic Executive (GOE). To supplement the team, the GOE recruited a small number of individuals for key roles with a proven track record of experience in delivering major projects. A delivery body, the Olympic Delivery Authority (ODA) was set up to deliver the infrastructure and venues. ODA was the contracting authority and appointed a delivery partner (CLM) to manage and administer the procurement and the delivery of works contracts. The key for both organisations was ensuring the right skill sets corporately and individually.

Significantly, the GOE was given authority to put in place a governance structure and processes that applied adequate formality to decision making and funding control while maintaining short decision-making timetables. The forum for key decisions and funding decisions and the release of funding was the Olympic Programme Review Group (OPRG), which was made up of representatives from the government funding and public body funding organisations, chaired by GOE. OPRG reviewed monthly schedule and cost positions and applied clear procedures for release of contingency funds, leading eventually to Ministerial approval. Regular reporting and review ensured that cost management, risk management and management of the limited contingency fund were aligned.

In 2008, the GOE pulled together a **comprehensive and realistic budget** for the Olympics, which included a contingency element. This budget was articulated in a readily available document, which also outlined the scope, scope ownership and time element for each element of the programme. On agreement of the budget, the GOE was given the authority to manage the release and spending of the budget under the governance of OPRG. This enabled fast decision-making, acknowledging the immovable deadline of the Olympics. Through its effectiveness, OPRG gained the confidence of government funding departments, which was the key to the decision-making process. Release of funding for projects was based on submission of business cases. Once approved, the ODA were given authority to manage the budget release and had delegated authority for release of contingency funds.

In response to the governance structure put in place, the ODA applied **effective procurement measures** to all the projects. They developed a transparent procurement code using the New Engineering Contract V3 (NEC3) standard form of contract. The adoption of the NEC 'cost-reimbursed' form of contract in the main was the key to ensuring flexibility and agility to overcome challenges presented by the project delivery. Acknowledging the disruptive nature of protracted disputes, the procurement code outlined a three-tier dispute avoidance and resolution process, which involved setting up an Independent Dispute Avoidance Panel (IDAP), a dedicated adjudication panel, with the final tribunal being the Technology and Construction Court.

At the outset **effective progress monitoring and risk management** arrangements were considered critical tools to manage the Olympics programme. Coupled with the use of 'cost-reimbursed' contracts, transparency of facts was the foundation that supported the governance structure and the timely decision making that was so necessary. This transparency through reporting was also made available to the public on a quarterly basis. This had a powerful effect of demonstrating a 'no surprise' approach and the opportunity for public scrutiny from which public confidence in the delivery of the programme was to evolve. Managing risk was put central to the decision-making process, with regular evaluation and adjustments made to the measures needed to minimise or ideally eliminate risks.

The final and probably key success factor was having people with the right skills to deliver throughout all the organisations. The Government acknowledged that it needed to act and perform as an 'intelligent' but lean client organisation. To this end, it recruited individuals from the private sector with a proven track record in delivering major projects. The fundamental shift that also occurred from traditional government procurement was the delegation and authority given to the individuals to ensure delivery to the budget and timescales within the governance process specifically developed for the Olympics.

In summary, the success of Olympic Park and the Games was underpinned by using a purpose-built delivery model with governance clarity. Central to this was the dedicated GOE team, led by specially recruited staff. The delivery organisations also recruited specialist individuals. For senior roles, acknowledgement of remuneration packages above background levels was necessary to attract the appropriate individuals with the right focus and behaviours. The challenge of an immovable deadline, whilst posing a significant challenge, also presented delivery opportunities which supported the need for timely decision making processes and delegation of authority to the appropriate people at the right levels in the organisations, because these people best understood the risks and the actions needed to manage these risks and to make the right decisions to meet the budget and time challenges. The model adopted allowed the ODA to realise savings and for the GOE to redistribute funds to cover other areas of risk pressures that developed in the operational readiness phase of the Olympics, such as the well-documented security staff aspects. The approach adopted by the UK Government is a testament to the boldness of the Government in changing its approach, with the result being world-acclaimed success.

All organisations were aligned to a common goal and the leadership came together regularly to evaluate progress and resolve issues. The common goal fostered a culture that meant there was only one possible outcome – hosting a successful Olympic Games. This ethos resulted in everyone associated with the programme taking personal responsibility and pride in their part in working towards that goal. Flowing from this, were positive attitudes that all issues needed to be resolved in a timely way. This was very evident in the final 12 months of preparation for the Games, when interfaces were at the most crucial, the GOE allocated a substantial element of contingency in order to resolve issues rapidly, coupled with this was specific delegation to a group to administer the budget on behalf of the Minister for the Olympics. This group dealt with up to 30 issues on a weekly basis making sound, but rapid decisions, which resulted in solutions being quickly implemented and excellent working relationships being maintained and even strengthened. This process continued into the operation of the Games and proved effective and contributed to the incident free aspect of the Games.

In conclusion, whilst the government and the supporting organisations set-up sound, practical governance and processes, the underlying factor was having the right people, with the right behaviours, doing the right thing at the right time. People, people, people! Source: National Audit Office, The London 2012 Olympic Games and Paralympic Games: Post Games review, HC 794, session 2012–2013, 5 December 2012.

23.3 Managing Risk – Tunnels for Heathrow's Terminal 5 (2001–2005)

Ian Williams MSc, CEng, FICE, FCIArb

23.3.1 Introduction – 1994 Onwards, the Years that Influenced how Terminal 5 would be Delivered

On 21 October 1994, a major collapse associated with tunnelling works occurred on the Heathrow Express project being delivered by BAA (owner of Heathrow airport). Fortunately nobody was killed but the incident resulted in prosecutions. A positive consequence of the collapse was that it was a catalyst for change in behaviour, attitude, and approach within the UK construction industry. The Health and Safety Executive's (HSE) final report (2000) on the collapse stated:

> 'those involved in projects with the potential for major accidents should ensure they have in place the culture, commitment, competence and health and safety management systems to secure the effective control of risk and the safe conclusion of the work.'

The scale of the collapse, its impact on the project and the negative impact on BAA's business reputation forced them and their suppliers to address the key lessons from the collapse. The fundamental factor to the recovery solution in 1994 was the adoption of a fully integrated team approach combining risk management with technical innovation.

The collapse also directly affected other contemporary tunnelling projects in London using the same tunnelling techniques. Within 24 hours of the Heathrow collapse, the owner, London Underground Limited, suspended two contracts on the Jubilee Line Extension (JLE) project using the same tunnelling technique. Over the next three months, the JLE project team carried out comprehensive reviews of the technical and management managerial approach of the works and approached the situation from a position of risk mitigation and management. The proposals were submitted to the HSE. After detailed review and evaluation by the HSE the projects were recommenced about three months after the initial stop. This was probably the first time a tunnelling project in the UK had used risk evaluation and management as a tool to evaluate construction decision making.

In the early 2000s, BAA embarked on its largest, most ambitious project and also one of the biggest projects in the UK at the time, Heathrow Airport Terminal 5 (T5). The project consisted of a terminal building and two satellites, 60 aircraft stands, car parking for 5000 vehicles, and 14 km of tunnelling to provide road and underground rail links to the terminal (Figures 23.1 and 23.2). The majority of the tunnels were constructed using tunnel boring machines (TBMs) and innovative developments of shotcrete lined method, Lasershell™ used for the construction of the numerous short, complex underground structures, such as shaft and cross passage complexes, (Williams 2008; Williams et al. 2004). T5 was the first major project that BAA was embarking on since the Heathrow Express project in the mid 1990s. Learning from the progressive changes in the approach to British project delivery in the late 1990s flowed out of the Latham and Egan reports and its own learning experiences from the Heathrow Express

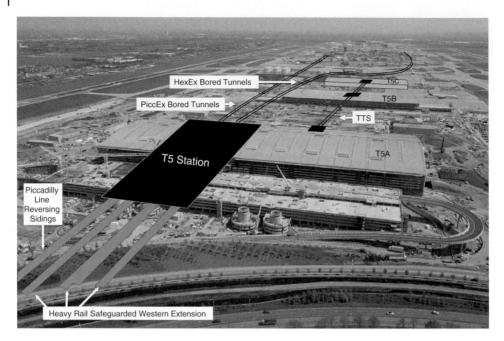

Figure 23.1 Aerial view of the T5 project (Williams 2008).

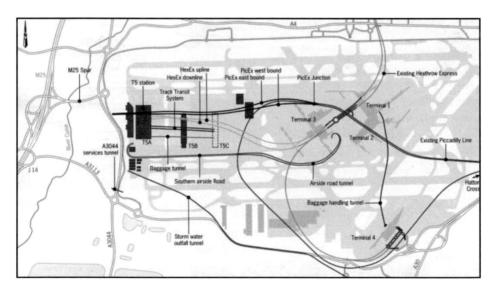

Figure 23.2 T5 project and the location of the tunnel projects (Williams 2008).

post collapse project recovery, BAA concluded that in order to meet the challenges to deliver the T5 project safely, on time and within affordability, it needed to develop a procurement and project delivery strategy that was able to meet the challenges of delivering a complex project. BAA accepted that to achieve its objectives, it was necessary to do something different to what was historically done by the construction industry. Taking on its learning from the Heathrow Express post collapse recovery and developments in the industry highlighted in the Latham and Egan reports, BAA considered the key success criteria for the strategy would be a bespoke, non-adversarial contractual approach based on relationships and behaviours with cost reimbursement as a cornerstone rather than conventional contractual approaches (Wolstenholme et al. 2008). BAA also concluded that all risks could not be eliminated but risk identification and proactive risk mitigation management could decrease the probability of occurrence and reduce the severity of the risk event with robust control processes and mitigation. Furthermore, BAA accepted that the risk of not achieving a successful T5 rested almost totally with BAA, which led to BAA accepting that it owned all the risks all the time.

Unconnected, but running in parallel with the early stage of the T5 project, the insurance industry dealing with insurance for tunnel construction projects noted that disasters associated with tunnelling projects across the world were becoming more frequent, with one of the notable ones in the UK being the collapse on the Heathrow Express project in 1994. Many insurers were considering withdrawal from the market leading to a potentially loss of confidence by client financiers which would have been disastrous for the tunnelling industry. In October 2001, the Association of British Insurers (ABI), representing insurers and re-insurers from the London-based insurance market, contacted the British Tunnelling Society (BTS) to share their growing concerns about recent losses associated with tunnelling works both in the UK and overseas. Rather than withdrawing completely from providing insurance cover for tunnelling projects, restricting cover and increasing premiums as a way forward, the ABI sought to work with 'industry' to develop a 'joint code of practice' for better management of the risks associated with tunnelling works.

In 2003, 'The Joint Code of Practice for Risk Management of Tunnel Works in the UK' was launched. The objective of the Code was to promote and secure best practice for the minimising and management of risks associated with the design and construction of 'Tunnel Works' (Clause 1.1 of the Code) as well and in the event of a claim, minimising its size. Furthermore, it provided insurers with a better understanding of the risks during the underwriting process and hence increased certainty on financial exposure in the provision of insurance cover. It was acknowledged, however, that strict implementation of and compliance with the Code would not necessarily prevent claims occurring but would reduce their frequency and severity.

BAA concluded that its approach and principles of the Code were aligned.

23.3.2   Terminal 5 (T5) Tunnel Projects – Managing Project Delivery Risks

Central to the success of the T5 tunnelling projects was the approach and strategies adopted to manage the delivery risks. The fundamental aspects were the form of contract (T5 Agreement), assembling the right team with the right values and behaviours, and the approach to risk management, which included the comprehensive adoption of 'The Joint Code of Practice for Risk Management of Tunnel Works in the UK'.

23.3.3   The T5 Agreement (the Contract)

The T5 Agreement (the contract) was a unique contract because it outlined processes and strategies to manage successful project outcomes in an uncertain environment.

It did not outline contractual positions for when things went wrong but focused on principles such as:

- People with the right values and behaviours
- People as individuals working on relationships
- Risk management
- Suppliers working in integrated teams to achieve a common goal
- Suppliers' actual costs were reimbursed with a guaranteed element of profit and an incentive target payment to reward exceptional team performance. In return, BAA expected World Class performance.

The historical consequences of cost and time risks embedded in traditional contracts were not shared, they were held by BAA. BAA owned the risks but expected their suppliers, in integrated teams to manage them.

Underpinning the Agreement were three fundamental working principles considered essential to make the Agreement work; trust, commitment, and teamwork. Emphasis was on selecting key people who had the right values and behaviours and coaching the delivery teams in the values and behaviours.

Figure 23.3 summarises the type of behaviours that characterise these values. For example the project needed people who would do what they say, focusing on outcome not problems, respecting the workforce and selecting people on merit.

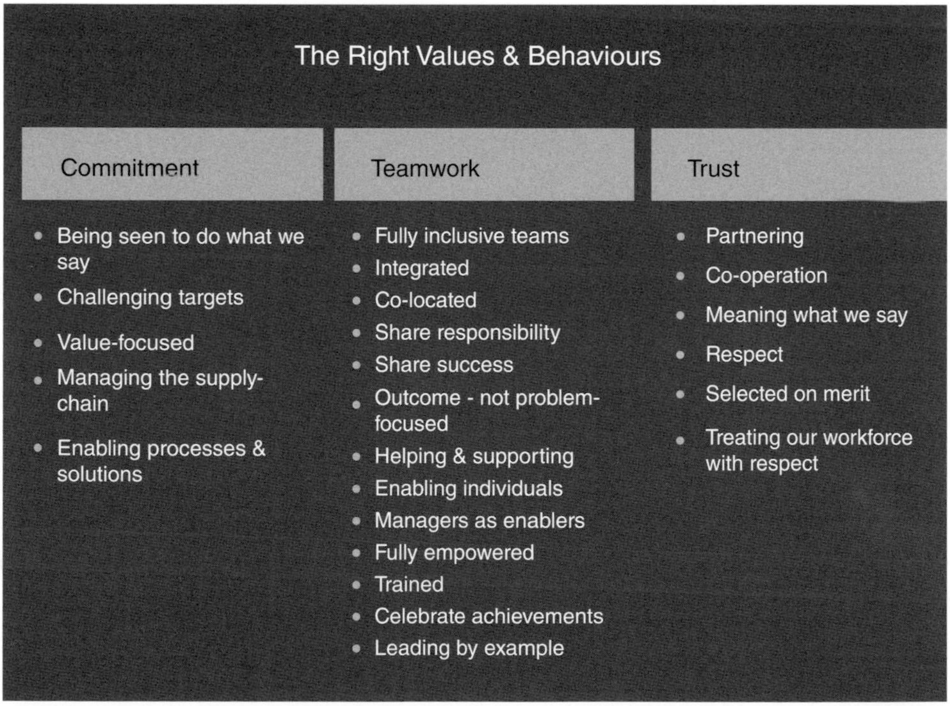

Figure 23.3 The right values and behaviours.

23.3.4 Project Delivery Team – Management Approach

Critical to the effective working atmosphere was creating a co-located single, multi-company integrated team and forming a project identity that displaced company identities. The division of the team was by function and role not by company. For the tunnelling works the team members consisted of BAA (client and project leadership), Morgan Vinci (tunnel constructors), Mott Macdonald (civil and tunnel designer), and Beton-Und Moniebau (sprayed concrete lining [SCL] design and on-site specialist support). The suppliers were selected on merit not on cost.

Specific roles and responsibilities were established and the team members were selected on technical track record and, above all else, the ability to work within an integrated team to achieve common goals. The key team members needed to be able to encourage throughout the team the values and behaviours outlined in Figure 23.3. Selecting the right team, coupled with the T5 Agreement meant that, above all, world-class engineers co-located in integrated teams could focus on engineering challenges and finding solutions, not traditional contract management to protect corporate contractual positions.

Apart from the usual project management/administration aspects, the key aspect related to tunnelling risks identified was the responsibilities of the designer and constructor and, critically, the independent role of checking the integrity of the design and the construction operation coupled with ground movement monitoring. The team that undertook this activity was part of the overall integrated team but was independent in function from the construction team. The team had direct accountability and reporting line to the client.

23.3.5 Workforce Safety – Incident and Injury Free Programme

Part way through the T5 programme, the leadership felt that to continue to achieve its objective of zero injuries, the project needed to make a shift in its approach to safety management. Looking to other industry learning, BAA concluded that it had to adopt a behavioural approach to safety management and committed to making this change from the workforce through to the programme leadership. A behavioural safety initiative was introduced, the Incident and Injury Free programme (IIF). The philosophy was getting safety to be personal rather than a routine following of process and procedure. The behaviours of a change in safety culture to influence safety performance were totally aligned to the T5 three values: Commitment, Teamwork and Trust. The implementation of the IIF programme is described by Evans (2008). Many of the suppliers adopted IIF or its principles as part of their company approach and carried through to later projects such as a number of the 2012 London Olympics infrastructure projects

23.3.6 Risk Management

Central to the T5 Agreement was risk management. The strategy of managing risk was through integrating risk and delivery management. The process involved initially exposing the risks through project risk reviews and thereafter regularly updating and reviewing 'live' risk registers and 'live' risk response plans within the framework of a project execution plan. Notwithstanding the approach to risk management as outlined in UK legislation, the Project Team made an early decision for the rail tunnels to use the guidelines of 'The Joint Code of Practice for Risk Management of Tunnel Works in the UK' (ABI and BTS 2003). It was understood that the T5 project was the first in the UK to fully adopt the 'Code'. Furthermore, by adopting an 'open door' policy for the insurer's advisor in the project team activities meant that the insurer was also part of the 'extended' integrated team.

On a project working level, IIF and risk management came together, and several innovations and initiatives were developed and implemented for the T5 tunnelling project. Adoption of the cause control measures identified by the risk management process and targeting attitude to personal safety were the two key elements of the strategy used to manage risks related to safety.

The philosophy of the team was to ask questions, bringing in representatives from the delivery team, managers, and workforce to discuss and establish the right approach to eliminate, and if not possible, to manage the risks. This took the form of regular office based workshops focusing on the immediate work activities in hand followed up by on-site sessions. This meant that the whole delivery team was involved in the solutions not just management as historically done.

23.3.7 The Outcome?

Fourteen kilometres of incident free tunnel construction was completed for the T5 programme without any notable effects on the infrastructures above. This has been documented in numerous publications.

The cornerstone of the success of the T5 tunnelling projects and the wider T5 programme was the contract BAA developed and implemented, the T5 Agreement, and BAA's acceptance that the risks and their consequences rested with BAA. The principles of the T5 Agreement focused on all parties being successful by aligned objectives and using risk management as a tool to support decision making to achieve programme success, enhancing workforce safety and reducing the likelihood and impact of unforeseen events. A fundamental element was the way the T5 Agreement was implemented, this being the three values and behaviours that underpinned the Agreement: Trust, Commitment and Teamwork. Part way through the programme, the IIF safety programme was introduced, which had a massive impact, much wider than just safety, on how the programme was delivered.

Without doubt, the fundamental factor underpinning everything was people. People who embraced the three pillars of the T5 behaviours: Teamwork, Trust and Commitment. Personally, and I am sure others who were also involved with the T5 programme, will look back and conclude that the experience gained was unique and would be a legacy benefit to next projects.

Final thoughts on future projects: many projects have the right contract, commitment to risk management and robust processes, and they have the right people with competence. However, their impact will be limited unless they have the right behaviours and values and can be relied upon consistently to do the right thing.

23.3.8 What Are the Key Lessons Learnt from the Above Projects?

Whilst these two projects are significantly different in terms of scope there are five key elements common to both and perhaps can serve as learning points other projects should consider.

- *People.* Both programmes realised that having the right people with the right skills was essential to achieve the objectives. People empowered and taking ownership for their actions.
- *Client ownership and appropriate procurement processes.* The client, BAA for the Terminal 5 programme and the UK Government for the London Olympics, took their role seriously and understood that failure was not an option. This led to how the project delivery was set up and the delivery models adopted, which were customer focused and acknowledged the commercial needs of the supporting organisations.
- *Governance and integration.* Establishing governance clarity was essential and acknowledging that to achieve the huge challenges, both programmes embraced and promoted the integration of many other organisations and stakeholders working towards an agreed single overall objective, be it a successful Olympics or a world class railway infrastructure.
- *Realistic budgets.* Budgets established by the people who were responsible for the delivery – ownership.
- *Management of risk.* Both projects adopted an open dialogue on risk and the challenges ahead, put managing risk central to the project and accepted that ultimately the ownership of the risks lay with the sponsor or client, but the management of the risk was owned by the organisations best placed to manage them.

The rest is history!

Acknowledgements

I would like to thank the many individuals from numerous companies and organisations that have been committed to improving the British tunnelling industry over the last 20 years. I would also thank the individuals who were fully committed to the principles of the T5 Agreement of Trust, Commitment and Teamwork, without it none of the successes at T5 would have been achieved. Notwithstanding the commitment of individuals, I also acknowledge the courage of BAA as a client to implement the T5 Agreement and being forthright in demonstrating client best practice and harnessing the lessons of the 1990s. Finally, I would like to thank Terry Mellors (insurer's representative) for his candour and invaluable support during the project and invaluable discussion and contribution of some of the material in preparation of this section.

Ian Williams

Bibliography

Egan Report. (1998). Rethinking Construction, The Construction Task Force, Chairman Sir John Egan, Rethinking Construction, DETR, London.

Evans, M. (2008). Heathrow Terminal 5: health and safety leadership. *Proceedings of the ICE* 161 (5): 16–20.

The Association of British Insurers & the British Tunnelling Society (2003). *A Joint Code of Practice for the Procurement, Design and Construction of Tunnels and Associated Underground Structures in the UK*. London: Thomas Telford publishing.

The collapse of NATM tunnels at Heathrow Airport. A report on the investigation by the Health and Safety Executive into the collapse of New Austrian Tunnelling Method (NATM) tunnels at the Central Terminal Area of Heathrow Airport on 20/21 October 1994. HSE. Books. 2000.

The Latham Report – Constructing the Team, 1994.

Williams, O.I. (2008). Heathrow Terminal 5 – Tunnelled underground infrastructure. *Proceedings of the ICE* 161 (5): 30–37.

Williams I, Neumann C, Jager J & Falkner L. 2004. Innovative Shotcrete – tunnelling for the new airport terminal, T5, in London. 53rd Geomechanics Colloquy and Austrian Tunnelling Day, Salzburg, Austria, 2004.

Wolstenholme, A., Ian Fugeman, I., and Hammond, F. (2008). Heathrow Terminal 5: delivery strategy. *Proceedings of the ICE* 161 (5): 10–15.

23.4 Cyber Design Development – Alder Hey Institute in the Park, UK

Stephen Warburton, RIBA

Director, Design & Project Services Ltd

A project won by Hopkins Architects in an international design competition in 2013, the £24 million Institute in the Park at Alder Hey Children's Hospital in Liverpool houses around 100 education, research and clinical staff. It is an important building as it is a centre of national excellence for research into children's musculoskeletal diseases and is the UK's only Experimental Arthritis Treatment Centre for Children.

As well as its main sponsors, Alder Hey Children's Hospital and the University of Liverpool, funding for the scheme was from a variety of sources, including the European Regional Development Fund, Liverpool City Council, and The Wolfson Foundation.

My involvement started in March 2014 where the challenge was to submit a tender that hit the brief in terms of cost and programme, whilst focusing on our teams skills to deliver a building of high architectural quality. Many years before, as a trainee architect in London during the summer evenings after work, I would walk across Regents Park and watch a few overs of cricket in the Mound Stand at Lords that Michael Hopkins had designed in 1987. Along with Foster and Rodgers, he had always been one of the 'greats' in my eyes.

I made a play on using 'cyber design development' or a 'virtual design workshop' as part of the tender submission, given the logistics of the wider team based in Bristol,

Liverpool, Manchester, and London, and the extensive travel that the design team would have to undertake on a regular basis to meet some tight preconstruction milestone dates. It proved to be unintentionally prescient, as I had imagined all project and design work-shops and meetings would naturally take place face to face in Liverpool, but more of this later.

So, after successfully negotiating the tender submission and some tricky interviews our team was announced as preferred bidder in April 2014. Straight away I knew this process would not be a typical run through tried and tested design management pro-cedures, supplier engagement and the obligatory value engineering process in the run up to contract award. No, I was on the train to Woking the next day to visit Hopkin's recently completed World Wildlife Fund headquarters with the project architect. This 'precedent' building was immaculately detailed in larch, western red cedar and the best exposed architectural concrete I had seen. This was what we 'had' to achieve in Liver-pool, no compromises, no short-cuts, no contractor design, just the architects vision delivered to perfection.

We set to work in May 2014, initially having our design meetings at the architects beautiful modernist offices near Marylebone in London then moving to the site in Liv-erpool in July 2014 after contract award. All was progressing well, we engaged David Bennett, a world renowned specialist and author of a dozen or so books on architec-tural concrete to rewrite the engineers' concrete specification and guide us in selecting the very best aggregates and just the right GGBS mixes to achieve our exposed concrete soffits and walls. Various grades of Scandinavian, Canadian, and Scottish western red cedar were investigated before selecting the one that would wrap around the building and fly across the glazing to create a seamless brise-soleil.

We were slightly ahead of the design delivery programme by the first week in August 2014 when I received a call from one of the partners at Hopkins saying that the project architect had been involved in a serious cycling accident in the middle of London and was having surgery in hospital (he had been hit by a lorry and had his spleen removed, I subsequently found out).

We carried on with the partner architect standing in on his weekly visits to Liverpool, but there was a problem. He hadn't designed the building from scratch, so no rough pencil doodles to basic cardboard model to CAD 'concept' model. No intrinsic under-standing of how sunlight filtered into the airy public areas and how the laboratories were cocooned within more environmentally moderated areas, or how the bespoke detailing should be resolved to get the best from the quality materials on show. That was all hours and hours of investigation and design optioneering that the project architect had gone through to create a solution or 'architectural vision' that the client wanted, but was now lost to the team as a whole.

A month later the impact of delayed design approvals and key areas still 'in abeyance' was starting to show, with the piling now complete and the concrete contractor already on site. The project architect was out of hospital by early September but under strict instructions from his doctor not to do more than two days a week in the office, and under no circumstances travel to Liverpool.

I had trialled WebEx earlier that year, so understood its functionality and how it could be used as a video conferencing facility from laptops or a boardroom from pretty much anywhere. Digging a bit deeper, it had various plugins that would allow all meeting participants to see real time drawing markups in Autocad or Revit whilst still seeing 'windowed' video and audio. After a few teething problems the system was working perfectly, the engineer in Bristol, architect in London and me in Liverpool flashing up photos or videos in HD of the site works taken sometimes only minutes before. It was a valuable tool, limited sometimes by never being as good as seeing mockups and samples first hand (although these were couriered down to London where possible), but allowing detailed discussion with as much visual material as was necessary to make informed decisions on whether a detail would work, or if it was back to the cyber drawing board.

Almost bizarrely, the WebEx design workshops proved to be better than the real thing due to the focus on a drawing by drawing, detail by detail basis that you naturally have to assume when using this format. No wandering off topic and discussing the performance of the client's project manager or other peripheral issues. It lent a rigour to systematically ticking off all the fine details that we had struggled with up to that point. It made me realise how little we get drawings out on the table these days as a collective design team and collaborate on finding the best solution through negotiating the pros and cons until a general consensus is found.

Many of the well publicised project failures in recent times stem from this lack of detailed investigation and rigorous 'testing' at the design stage. At Alder Hey we demanded that all package interfaces were understood explicitly so any risk in subcontractor to subcontractor design coordination had been mitigated well before they arrived on site. Through the WebEx process we could Google and share specific technical data from specialist supplier's websites as we were discussing a problem sitting in different parts of the UK. The sophistication of the major manufacturers' and suppliers' online specification portals is now so advanced that 2D and 3D details are nearly always immediately available, backed up with full specification, warranties, test data, BIM modelling information and environmental accreditation.

In summary:

- Immediate shared access to online specialist design information suits rapid decision making and provides quicker high-level coordination.
- An online design agenda should be carefully tailored to focus on specific outputs, so less on design team progress, more on visually reviewing and resolving.
- Due to the focus required by online participants, shorter more frequent sessions get better results.
- Any perceived 'downtime' by design consultants is easily offset by the lack of any travelling to a live meeting.
- It can be difficult to minute interactive online meetings, so (with the participants consent) WebEx sessions can be recorded and linked to key actions for distribution.

23.5   The Importance of Clear Ownership and Leadership by the Senior Management of the Client and the Contractor

Charles O'Neil

In all development and construction projects the prime objective of both the client and the contractor should be to complete the project with excellence in all the main components of delivery. These are essentially the same for both parties and include:

- A facility that meets or exceeds all the functional and commercial feasibility requirements of the client.
- An aesthetically attractive design that is environmentally compliant and 'pleasing'.
- Delivery on time, on budget and to the specified quality standards.
- A satisfactory financial result for the contractor, design consultants, suppliers, and subcontractors.
- An unblemished safety record during construction and commissioning.
- Effective risk management processes that provide early warning of potential issues, thus enabling timely action to be taken to forestall or mitigate these issues.
- Managing and resolving conflicts in such a manner that these issues do not turn into formal disputes.
- Excellent working relationships with the relevant authorities and the community.
- Effective communication and relationship management protocols that contribute significantly to the delivery of the above objectives.

If the above objectives are all achieved then the end result will be the delivery of a first class project, *but this can only be achieved if there is clear project ownership and leadership evident from both the client and the contractor during all stages of the project.* If their performance is first class in these areas then this should flow through to the approval authorities, the design team, and all the suppliers, subcontractors, and stakeholders.

However, if there is a lack of ownership and leadership at senior level then the impact on the project can become quite detrimental to its future viability and operations; stakeholder relationships will be damaged; disputes will occur; and there will possibly be serious financial implications for some or all of the parties to the project.

What then are the key human factors that are required at senior management level of both the client and the contractor in order to ensure that the above deliverables are achieved?

- 'Project ownership'. This is the passion to deliver a project that all can be proud of; to accept full responsibility for decisions and actions taken and to *'make it happen'*.
- Mutual trust and respect at senior level of the other stakeholders and their objectives.
- Sound decision-making, taking into account the needs of the project and of the other stakeholders.
- Must be a strong advocate of a collaborative approach.
- Lead from the front with best practice project management, risk management and digital platforms, such as BIM.
- The ability to communicate clearly and unambiguously, including spending personal time with team members and the other stakeholders.
- Natural leadership and people skills that motivate and create team spirit, loyalty, and a strong desire amongst the team to achieve the project objectives.
- Being able to select the right team, having trust in them and delegating responsibility.
- Thorough planning and efficient implementation are essential components of success, and likewise experience, anticipation and the ability to see the bigger picture are essential requirements of a good leader.
- Effective leaders are good delegators and keep some time up their sleeve for strategic thinking, planning, and public relations.

- Sledgehammer managers, yellers-and-screamers and bullies might meet the programme and get the job built, but they will never engender team spirit and loyalty or gain the respect of team members or the other stakeholders.
- Successful leaders invariably couple their experience with astute anticipation and effective risk management, and then make full use of good communications and relationship management to resolve issues and keep the project on track.

Some people might say that it is too ambitious and not realistic to strive for and hope to achieve the excellence in delivery that I have described, however I can assure readers that it is entirely possible. I have been proud to participate in a number of projects that have met these goals, including the following quite diverse ones:

- The Mersey Gateway Bridge, UK
- Kelowna and Vernon Hospitals, British Columbia, Canada
- M80 Motorway, Glasgow, Scotland
- Golden Ears Bridge, Vancouver, Canada
- Royal Women's Hospital, Melbourne, Australia
- Somerset Chancellor Court, Ho Chi Minh City, Vietnam
- Moree 125 000 t Grain Storage, NSW, Australia

We all understand that it is rarely a perfect world when it comes to delivering infrastructure developments and construction projects and it is quite common to have hiccups with finances, procurement, authority approvals, subcontractors and the like.

However if a project is soundly based in terms of there being a 'need' for the facility and the commercial feasibility is viable then these hiccups should be viewed as issues to be resolved in a positive manner.

The difference lies in how you resolve them and this is where clear ownership and leadership on the part of the client and the contractor has great importance. If the senior management of these parties and other stakeholders in the project all have a common objective of delivering excellence then these issues can invariably be overcome and disputes avoided.

Strong leadership will create a project spirit wherein nobody wants to 'spoil the party'. A perfect project is when outsiders say 'weren't they lucky to have such a smooth run' – but it has nothing to do with luck!

Section E – Robust Processes – Corporate and Project Management

24

Planning and Programming Major Projects

Charles O'Neil and Rob Horne, Partner at Osborne Clarke, UK

24.1   The Foundations of Success

Surprisingly few people are really good at planning and programming major projects. It is a specialised skill and it is one of the most important professions in the development, construction, and engineering industries. It is the foundation of success for all property developments, traditional construction, D&C contracts, and Public Private Partnerships (PPPs).

Contractors, consultants, and their clients can choose from several proprietary planning and programming formats, or use these as the basis to develop their own internal ones. Commonly used formats include:

- Gantt charts, commonly referred to as Bar Charts (see Figure 20.1)
- S-Curves, including Earned Value Analysis
- Time location diagrams (also known as 'Time Chainage' and 'Time Distance' diagrams).

S-Curves, Earned Value Analysis and Time Location Diagrams are explained in more detail in Section 24.2.

The key to good planning and programming is being able to keep your eye on the bigger picture. The starting point should be the Master Plan Summary sheet and this should have no more than 20 activities on it, with its own simple Critical Path line (it works well on a bar chart). It should literally be one sheet of paper, maximum A3 with sufficient notes and explanation of the activities (in a legible font size) that the overall nature and sequence of the works is obvious. The Master Plan Summary should not be just a group of rolled up activity bars that are drawn from the underlying detailed programme, but created independently as a 'bigger picture' senior management tool. It is a very important overview.

No matter how big a project is, when you really examine the bigger picture there will not be any more than approximately 20 main activities that are critical to the overall delivery programme. They may change during the duration of the project but the main critical activities will still be small in number. This is the sort of Master Plan Summary that senior management of clients and contractors want to look at, not several pages with 500 activities.

Global Construction Success, First Edition. Charles O'Neil.
© 2019 John Wiley & Sons Ltd. Published 2019 by John Wiley & Sons Ltd.

Behind each of the main activities there may be up to another 100 activities, which can be produced on back-up sheets, again with their own critical path line. The Master Plan Summary and the multiple activity programmes are all produced on Bar Charts.

A common mistake made by some programmers is to produce all the activities for a project, several hundred of them, across a few A3 sheets, all beautifully in sequence and with coloured critical path lines all through them, but without a Summary sheet.

The problem with this is that:

- You can't see the big picture easily.
- The programme is pretty well useless as a management tool for section supervisors on construction sites. It is simply too much to look at every couple of days, as they should, and they will largely ignore it.
- However, if you give them a concise programme that only relates to their section of the work they will use it and give regular feedback to the programmer to update it. Of course they should also have a copy of the Master Plan Summary so they can see how their work fits into the bigger picture.
- For updating, this approach is also much better than having the programmer wandering around the site coming to his own conclusions on progress without talking to the section construction managers and section supervisors, which is not uncommon and quite impractical (yes, it really happens – another breakdown in communications).

When looking at a programme for a major project, with say a three to four year design and construction period, there are four things to immediately look for:

- *Master Plan Summary sheet*. Rarely available unless requested, but essential to give a quick snapshot of the project and key areas of risk.
- *Revision number and date*, also showing how many times the programme has been changed and updated, with the relevant dates.
- *Missing items*. Anything that is really basic to the delivery process that might be missing. It happens; people get too close to the project. It might be related to design or authority approvals, construction or commissioning. In the anecdote at the end of Chapter 4 'Project Leadership – How Bad Can It Get?' there is a classic example of something fundamental that was missing that had serious delay implications and should never have been missed.
- *Commissioning and Completion Activities*. Go to the last page of the programme and look at all the activities for commissioning and completion. If there is half a page of activities all bunched up into three or four months in the bottom right hand corner then your alarm bells should start ringing, because it is a sure sign of potential trouble. This is one of the most difficult and complex periods of any large construction project, yet this bunching up shows that the site management and the programmer have not given it much thought and invariably not enough time. This is a very common problem and is a **_high risk area_**. Appropriate time being allowed for this period is more important than absolute detail on every activity as testing and commissioning plans often develop during the project, particularly one of three to four years duration.

The following are a few practical tips for effective planning and programming:

- The 20-activities Master Plan Summary is a **_must_**.
- 'Whiteboard' the BIGGER picture and work backwards from the target date as the required start dates that are established can be surprising

- Do not let the automated software do the critical path. It is clever, but it doesn't think as well as you do or know about all the implications. Think through all the issues yourself and install the critical path accordingly. There is nothing wrong with using software like MS Project, in fact you definitely should, but you must carefully control the programming inputs and then do a common-sense check of the output. And yes, it is not unusual to hear someone say 'but that's how it came out of MS Project', which also shows up their inexperience. The old saying 'garbage in, garbage out' couldn't be more relevant.
- To get the right picture, try sitting back with your eyes shut and walk through the project in your mind and keep asking 'what has to happen to make this happen?' and 'what if it doesn't happen?' Visualising is a very effective technique.
- Don't be afraid to act like a child with constant, simplistic 'but why?' style questioning. Challenge assumptions, test everything. The power of 'but why?' is known to every parent in the world, just never accept an answer of 'because I said so'.
- Some of the best programmers in the world print out the A3 sheets and then attack them with their big red pencil and ruler.
- There is quite a bit of misconception about what a critical path really is. People tend to forget that a good contractor should be flexible in organising the works around unforeseen delays in say material and equipment deliveries and has an obligation to try and mitigate overall delay, for the sake of their own pocket as much as anything. This issue of misunderstanding commonly pops up in claims for Extensions of Time (EOT). It is a complex contractual area and is a study on its own.
- Programmes are sometimes designed to influence clients and lenders because of the cash flow implications (or pull the wool over their eyes with downright dishonesty). Keep well away from this practice.
- Another dangerous but not uncommon practice is 'head-in-the-sand' syndrome, where project managers have got themselves into a corner, can no longer accept reality and produce programmes to protect or project the reports they are submitting. If you suspect this, ask for details of the programme changes over the last couple of updates. Manipulation of the programme in this way often occurs by amending logic links, deletion of activities or insertion of non-working periods to 'force' the critical path to take a different route through the activities.

24.2 Monitoring 'Progress versus Programme' and 'Cost-to-Complete versus Budget'

24.2.1 S-Curves and Their Significance

S-Curves basically reflect common sense. The more effort you put into the planning and the preparation of the specifications and contractual documents upfront then the smoother will be your design and construction programme; with minimal issues, variation claims, delays, and cost problems. This applies to both clients and contractors.

S-Curves therefore are the very antithesis of so-called 'fast track' programmes, which in essence say 'let's start building tomorrow and we'll sort out the design and specifications as we go, on the run'. Over the last few decades it is rare that 'fast tracking' has worked well on a major project, but there have been many disastrous attempts, a few of which have been highlighted in this book.

However, do not confuse 'fast tracking' with 'high performance' contracting. The latter is all about careful and detailed planning, programming, and risk management compressed into a much faster than normal delivery period and may involve 24/7 operations; but all still in line with the principles of a sensible S-Curve.

Fast tracking on the other hand is just a 'start early (and we should) finish early' mentality without rigorous planning and programming and thorough consideration and challenge of all the risks and delivery issues. Invariably, the result is late completion and over budget.

With high performance contracts there will be a real reason for doing it this way and the costs will invariably be higher, but it is done with eyes open and with the implications known in advance, as with the following project.

Some years ago a starch factory blew up in Australia. It was a huge dust ignition explosion and it scattered parts of the factory over a wide area and destroyed much of it. At the time it was the only starch factory in the country and the owners had a virtual monopoly in the market; however another company had announced that they intended entering the market and had commenced planning their new factory.

The owners of the factory that blew up were determined to continue supplying their customers and immediately started flying in product from one of their overseas factories, obviously at huge cost.

The rebuild of the factory was project managed on a 'high performance' basis, with the project managers and subcontractors working hand-in-glove with the owners. It was a big task. The factory covered a whole city block and was six to eight levels high on average. It was in an inner city location so the façade had to be brick to comply with city planning and there were all sorts of other inner city location implications. The company wanted the latest technology and equipment, but the waiting list in Scandinavia was six months. They paid a premium to jump the queue and chartered Jumbos to fly it to Australia.

Two months was spent on the planning, design and authority approvals, working around the clock; with the demolition and early works happening simultaneously. The main construction started at the beginning of July and the new factory was commissioned and started production at the end of February, to everyone's amazement. It was the result of very detailed planning of every aspect.

In summary, an S-Curve demonstrates that the success of the project will be in direct proportion to the amount of forward planning, but if you cut this short then the top of the curve will most likely flatten out and result in higher costs and a longer delivery time.

24.2.2   S-Curves with Earned Value Analysis

It is common for S-Curves to be adapted to include Earned Value Analysis, which is essentially a cost loaded programme that gives the programmer accurately updated cash-flows with any changes to the works programme or costs. In addition to the vertical total costs that are shown, there can be horizontal cash flows aligned to the progressive programme that show costs-to-date and costs-to-complete versus budget, similar to what are often used with Gantt (Bar) charts. The information that can be shown on the S-Curve includes:

- Cost overruns or savings
- Schedule (programme) slippage or gains
- Time related cash flows – with these the most informative figures to include should be 'costs-to-date' and 'costs to complete' compared to a parallel line showing the original planned costs (budget) per month, which should never be changed.

Examples of S-Curves with Earned Value Analysis can be readily found on the internet and a typical S-Curve is shown in Figure 24.1.

24.2.3   The Importance of S-Curves to Property Developers and Their Lenders

The S-Curve progress has vital importance to property developers who are using a high proportion of financing and also with PPPs where the financing can be as high as 90% of the total capital cost.

With all commercial projects the completion date is the trigger for the operating revenue flow and any delay can have serious consequences on the return on investment, let alone the further downside if the project runs over budget. The lenders watch this like a hawk obviously, because it does not take a very large over-run in time and cost before a project can run into serious financial trouble.

The two main things to be watching are 'progress versus programme' and 'cost-to-complete versus budget' and these two areas can be quite accurately compared on a properly prepared S-Curve. It is important not to confuse 'cost-to-date' with 'cost-to-complete', because 'cost-to-date' is only a component of 'cost-to-complete', which must take into account all potential additional costs that will be incurred in getting to practical completion, with the cost of additional time and 'preliminaries' being just as important as the direct construction costs.

Time location diagrams (alternatively 'time chainage' or 'time distance') are commonly used on linear projects, such as railways, roads, and tunnels, that usually require the excavation and movement of large quantities of earth and rock.

They are different from the traditional Gantt bar chart program and S-Curves in that they provide a flow of visual data in terms of time and location on a two-dimensional plan that acts as a visual connection between the project master plan and the project worksite itself. The diagram provides real-time monitoring of 'progress versus planned' as the work teams and equipment move along the construction site. As an example, the diagram would show when the rock excavation is scheduled for the 6 km mark and the expected duration of the work. The diagrams can also include information on resources, quantities, and costs as well.

24.3   Extensions of Time, Concurrency and Associated Costs

This is one of the most complex areas in construction management. Claims for EOTs and costs arise in virtually all projects, certainly in all major projects, and commonly lead to disputes over how they are assessed. We will only give a brief summary of the overall picture because many books and papers have been and continue to be produced on this topic.

From a risk management point of view it is an extremely important topic in that both parties, the client and the contractor, are trying to protect themselves financially. The stakes are big, the assessments can be difficult, and the relevant terms of contract need to be clear and precise.

24.3.1   Guiding Principles for Delay, EOTs and Cost Reimbursement

- The lump sum price and the completion date are fixed in the contract at the time of signing to protect the client and safeguard their investment against costs incurred through contractor delays and claims. Having done this it is then necessary for

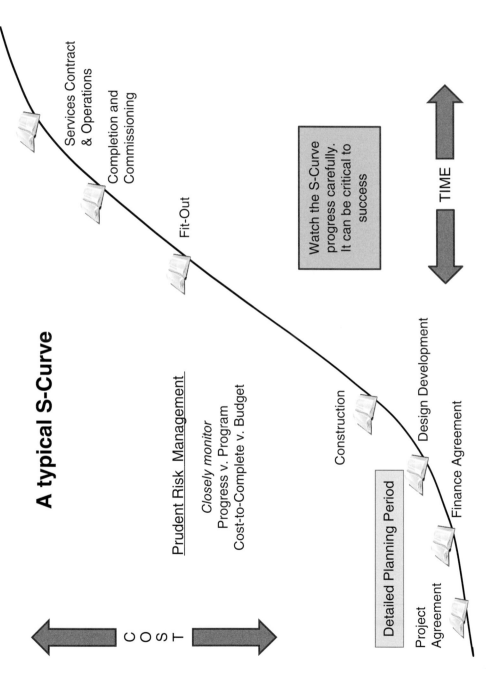

Figure 24.1 A typical S-Curve.

the client and their consultants, such as the architect and the project Engineer (independent supervisor), to undertake their administrative responsibilities and contractual obligations in a timely and efficient manner. If this is done and the contractor is late in delivery, then Liquidated Damages (LDs) and other such penalties as written into the Contract may be applied to the contractor. It is important to remember that LDs will be the client's only means of recovering time-related losses. In order to be effective they must relate to a specific completion date (or one that is identifiable with certainty) and a specific amount of damages. The calculation of the amount of damages should, in the UK, be such as to protect a legitimate business interest without being extravagant, exorbitant, or unconscionable.[1] A reasonable shorthand in many jurisdictions around the world, though not entirely correct in the UK, is that LDs should represent a 'genuine pre-estimate of loss'.

- On the other hand, the contractor is protected against LDs, penalties and costs arising out of client related delays by EOTs and related cost reimbursement. EOTs can result from any delay caused by third parties related to the client and not under the control of the contractor, such as directions issued by the architect and Engineer.

- If there is 'client-caused' delay, e.g. lack of instruction, and simultaneously a 'contractor-caused' delay covering part or all of the same time period, e.g. lack of site labour, then under established international practice, the 'client-caused' delay takes precedence over the 'contractor-caused' delay and the contractor is entitled to the EOT, because based on the 'prevention principle' it would not have been possible for the contractor to finish any earlier and therefore he should not be penalised, and the client cannot profit (via the impositions of LDs) from its own wrong in causing the delay. The question of recoverability of cost during the period when both delay events are running is equally difficult with no set single position. However, best practice is probably that while the contractor recovers time they do not recover money. That way a jointly caused event (or a concurrent event) is as close to neutral between the parties as is possible in the circumstances.

- In certain circumstances (and in some but not all jurisdictions[2]) it may be reasonable to give a proportional EOT that reflects the amount of delay attributed to each party, but only in circumstances that are quite clear in measurable terms, with any doubt being in favour of the contractor. Equally, where contractor costs can be applied exclusively to one delay or another they generally should be, whether recoverable or not.

- The Delay Event must be shown to have been on the Critical Path for an EOT to be granted, in other words it must have actually delayed the completion of the project. Delay off the critical path may give rise to recoverable cost but not time. If the specific works that initiate the EOT claim and the flow-on sequence of works could still have been done without delaying the Completion Date (which includes previously approved EOTs) then no EOT or overall time related costs shall be applicable (disruption and other noncritical time related costs may still be recovered however).

- Most contracts will require notice periods for (i) Notices to be given within a prescribed number of days of becoming aware of the delay event and (ii) the submission of substantiated claims within a prescribed number of days from the cessation of the

---

1 Cavendish Square versus Makdessi and Parking Eye versus Beavis, both UK Supreme Court decisions in 2016.

2 A proportional award is, for example, possible in Scotland but not England under the law current at the time of writing.

delay or earlier, with regular updates until the delay has ended. Contractors must make themselves aware of these periods or suffer possible time-barring of the claim.

- The contractor has an obligation to mitigate delay, if at all possible, by reorganising the site activities in order to maintain the Contract programme, as adjusted for any EOTs already approved. In other words, the contractor must show that the Delay Event in question did indeed cause an extension to the Completion Date and there was no way this could be avoided. What is required by way of mitigation is complex, but suffice to say here it should always be considered and should not give rise to additional cost.
- Where several Delay Events occur, the contractor is only entitled to the net extra time required to complete the construction. On this basis, several Delay Events may only result in one EOT being approved even though a similar amount of delay time may apply to all of them. In this situation only one lot of fixed site costs will also be approved, but specific costs related to individual delay events may be approved if the contractor can prove them. Looked at slightly differently, one is looking at 'waves on the beach'; it is the ultimate end point overall that is important albeit there may be numerous 'waves' covering the same distance.
- It is possible in some situations that the contractor will be granted an EOT but no costs, which will give them protection against potential LDs. This is commonly known as 'time without money'. As noted above, this is common where there is concurrency of contractor and client delaying events.
- The contractor must prove the specific losses claimed in relation to Delay Events and EOTs by providing in each case an itemised breakdown of costs, which is evaluated on a reasonable basis. The cause (delay) must be shown to link directly to the loss, commonly referred to as demonstrating causation or a causative link.
- A contractor might assert that delayed payments have caused slow progress on site or an inability to place orders for long lead time materials and equipment, but such an assertion should be assessed by comparing payments-to-date with the value of works-to-date, including off-site orders that have been placed.
- The contractor may be entitled to separate compensation for delayed payments, such as interest or financing costs, if this is written into the Contract, but an EOT will not necessarily be appropriate unless the delay in payment is for an unreasonable amount of time and can be shown to have caused a delay to the overall programme, e.g. the delayed payment caused a delay in procurement. In such circumstances, the entire financial situation needs to be taken into account, including any advance payment, which should give the contractor positive cash flow. It is quite rare, but some contracts include an automatic EOT with Costs built into the Contract, which comes into play if the client misses the contractual payment deadline, such as 56 days after certification by the Engineer. This is a strong incentive to the client to make the payments on time.

24.4 Ownership of Float

'**Ownership of float**' is a notoriously difficult subject, not least because there are multiple different ways of 'float' being defined and appearing in a programme. Subject to the wording of the contract, the float could be owned by the employer or the contractor or indeed the project. **Float is free time within a critically linked programme**. It is a gap or buffer along the critical path from the start to the contractual completion. It could be inserted intentionally (in which case it could also be referred to as a risk allowance to try and ensure the contractor retains ownership of it), be a function of the critical connections within the programme (where activity C is driven by both activity A and B, activity A is planned to finish on day 20 whereas activity B finishes on day 15 giving 5 days float at the end of activity B or 5 days of risk allowance depending on how the contractor drew up the programme).

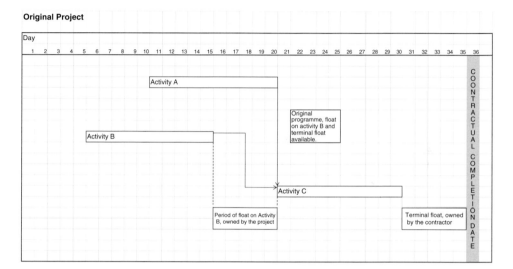

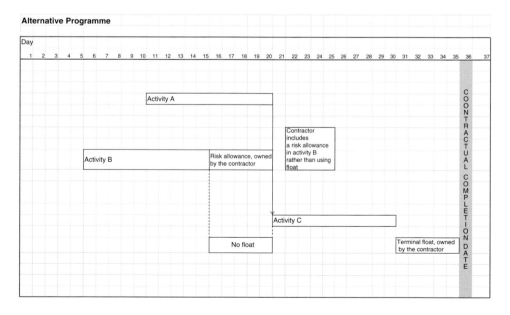

Or float can be a consequence of progress (i.e. an activity being completed more slowly than planned. Taking the above example, if activity B were delayed by the contractor by five days it would it would become a joint driver with activity A. Where there was float in the programme the contractor delay uses it; where there is risk allowance again the contractor will use it to mitigate the delay to the project.

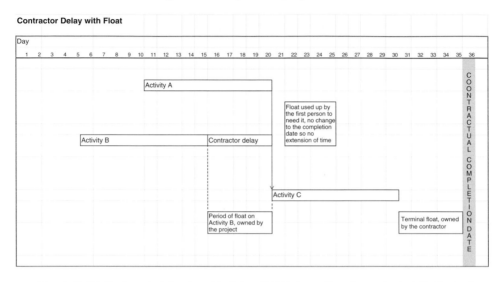

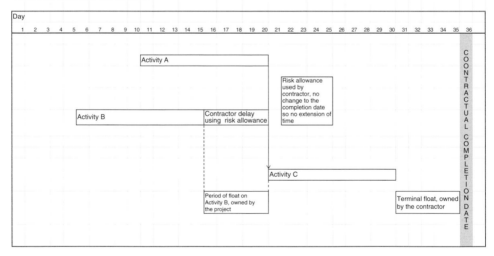

However, the difference between float and risk allowance is clear when there is an employer risk event. Where there is float there is no delay, the employer event using up the float. However, where there is a risk allowance, there is no float to use so the contractor will be entitled to an extension of time.

Employer delay with float

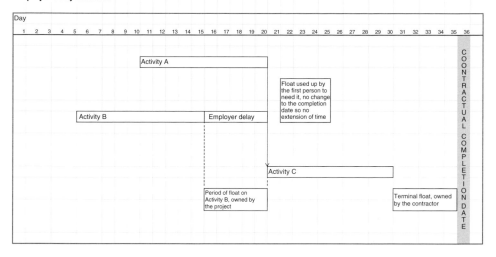

Employer delay with risk allowance

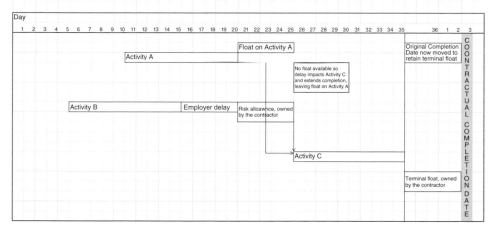

The float could be between activities (as in the examples given); within an activity (again probably better described as risk allowance); or at the end of the project where in effect the contractor is saying they intend to complete early (this is shown in the above examples as 'terminal float').

It is not uncommon in newer forms of contract for float to be assigned to one party or the other, but where it is assigned to the employer, in practice it will often simply be hidden by the contractor within activities to ensure they retain control of it. However, consequentially, the activity durations are not as certain as one cannot tell which, if any, contain risk allowances, giving a less genuine and clear programme and one which is harder to monitor progress against.

Where the contract is silent, it is most often understood that float is owned by the project, in other words, the first person to need it uses it; so if there is an employer delay of five days followed a week later by a contractor delay of five days and there was five days of float in the programme, the employer delay would not change the end date of the project whereas the contractor delay would, making the contractor liable for damages. If there had been a risk allowance rather than float the position would be reversed with the original employer delay changing the end date and giving the contractor some recovery whereas the contractor delay is absorbed through risk allowance and terminal float.

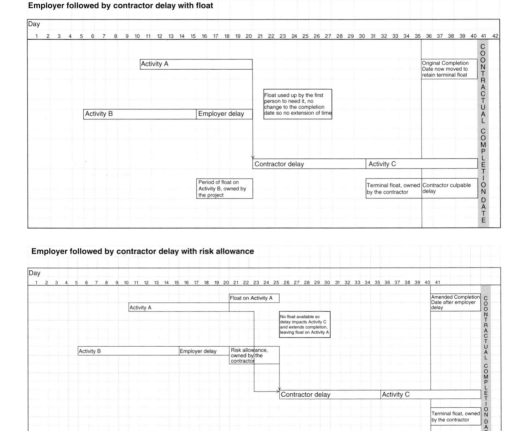

It is always a worthwhile exercise, when reviewing a complex programme, to consider not just activities with zero float (that is those forming or attaching to the critical path) but also those close to critical, usually less than five days of float. This is because a critical path can jump quite quickly and understanding where all of the pockets of float are on a programme as a whole, not just the current critical path, is essential to effective management.

25

Managing and Resolving Conflict
David Richbell, FCIArb (Mediation)

25.1 Conflict Can Be Good

Most people consider conflict to be something to avoid, a negative influence that can only cause pain and discomfort. In fact conflict can be a catalyst for understanding and innovation, and a force for progress. Disagreement can cause discussion, an exploration of the reasons for differences, that can lead to understanding, which can lead in turn to agreement and co-operation. It just needs people to resist the instinct for fight or flight or freeze and to pause and be prepared to talk. And listen. If this happens then it can lead to understanding, valuing differences and a preparedness to find a joint solution.

25.1.1 Different Truths

We are human beings; all different, all emotional, and all the product of our genes and our environment. We see the same facts and events through different eyes and process them differently. It doesn't necessarily mean that our version of the truth is any more right or wrong when set against someone else's truth – it is just different. Once we under-stand that, we should be able to find out why someone else's truth is different to ours, and to accept that their truth is genuinely held and should be valued alongside ours.

25.1.2 Difficult Conversations

Because most people want to avoid conflict, they will usually avoid having a difficult conversation. Sometimes the problem will go away – time may be enough for the issue to evaporate or sort itself out. Unfortunately that is only sometimes. Most times the problem will get worse because it has not been dealt with at the start until it explodes as a much more serious issue later on. People need to feel that they are being heard when they have something to say that is important to them. If they feel they are not being heard they raise their voice in the hope that someone will notice, but it usually has the opposite effect because people instinctively want to avoid others who are shouting or being difficult. The trouble is if that doesn't work the voice becomes more strident, and if that doesn't work they resort to violence. And violence breeds violence. Not good.

But it does take a real effort to address an issue face on at the start and not to let the instinct for fight/flight/freeze take over. However, making that effort can reap riches.

Just to enquire why a person sees the same facts and events differently, or why they feel so strongly about something, can make all the difference; just taking time to enter into a dialogue will in itself change the dynamic. People need to feel heard and demonstrating that you are listening and really hearing what is being said in a non-judgemental way will lead to a rewarding conversation.

25.2   Co-operation *Versus* Confrontation

We are in an age of fighting, of trying to win. Our culture is to fight for our rights, to seek justice through the courts, to oppose those who have different views.

The trouble is that lifetime habits are difficult to change. Western culture, or more particularly, British and American culture, is founded on positional negotiation. That invariably means building the best possible position and giving little, slowly. That usually either leads to deadlock and a failed negotiation, or to one party getting the better of the other. Both results almost always damage the relationship. However, co-operating to achieve the best outcome for all parties is, for most people, counter cultural – we like to 'win'. So setting aside the need to win, and working alongside others, can be quite a challenge. It involves a resistance to instinct and a stepping back from 'rights'. It means listening instead of demanding, exploring options instead of quarrelling, and seeking interests rather than fighting claims. It needs extra effort, but the outcome, and the experience of negotiating in a better way, can be life-changing.

25.2.1   Making the Effort to Co-Operate Is Only the Start

However, making the extra effort is only the start; there are other challenges on the way:

- We are all different
- Different nationalities view time, seniority, justice differently
- Power imbalance can be significant
- When trust is broken it is difficult to rebuild
- People don't necessarily play by the same rules
- Reluctant or absent parties are an issue
- Some people want to be told the answer
- Some people don't care about preserving relationships.

25.3   We Are All Different

Actually, our similarities are probably more basic and so more important. We want to be loved – and to have an outlet for our own love. We want to be respected and to retain our dignity. The need for food and shelter, warmth, and human contact are basic to us all. However, we are also unique; each one of us is different and so our emotions, our demeanour, our way of relating to others, of making decisions, our response to conflict (and to pleasure), all are unique. And all affect the way we communicate and negotiate. Add to that the cultural dimension, and we are in a potential minefield.

25.3.1 Cultural Differences

The first question is: 'what do we mean by culture?' The best answer to that question is 'it is the way we do it here'. It is so easy to assume that culture means ethnicity, custom, religion, tradition, and so on. That may well be so but it is so much more complex than that. Every community has its own culture and, unless you come from that community, there is a danger of being unaware of particular sensitivities that may have an impact on any communication or negotiation. Indeed, even being a member of that community is no guarantee of knowing all the sensitivities! We are all individuals with our own characteristics, and few of us conform to the assumed group characteristics. We may well conform in many ways to a group identity but no human being acts all the time in an individually predictable way. So the danger of such ignorance is that we make assumptions, usually based upon stereotypes, and assumptions are a weakness.

25.3.2 Changing Assumptions into Fact

Any position based upon assumptions is a weak position. It should always be questioned and then find ways that can change assumptions into facts. The fewer assumptions there are, the less risk, and therefore the stronger the position taken. Researching, finding other's experience of the individual or group, even enquiring of the individual or group itself, will provide more information and therefore build more confidence before entering into the communication or negotiation itself. So be open to learning that 'they' are not as expected. The wide variations among cultures do provide clues to follow up, but all of us have interests, qualities and personal preferences that do not fit a standard mould. It is called the richness of life! So the rule must be to turn assumptions into facts. It will lead to a far stronger, and therefore less risky, position.

25.3.3 Time, Seniority, Decision Making

Having said all that, there are some important cultural issues that need to be clarified before business can commence. Time is one of them. Most western cultures view time as critical and are used to deadlines and the need for closure so that the next issue can be addressed. Some cultures see time as endless and have a need to consider, discuss and gestate before a decision can be reached. This can be very frustrating to a deadline-controlled negotiator, but prior knowledge will allow the negotiator to plan a strategy that will use the time efficiently.

Similarly, the matter of authority could be an issue. In some cultures, particularly eastern and oriental, the most senior person often speaks the least. Indeed, the real decision maker may not even be present. In others, the final decision may need to be collective, so taken back to the group by its representatives before a solution can be ratified. Of course, that may well be present in western negotiations, but born of a reluctance to take responsibility in case of criticism (as witnessed in many UK Government and local authority negotiations). This underlines the importance of clarifying the situation beforehand. Who is the decision maker? Does a deal have to be ratified? What time scale is involved?

25.4 Fairness or Justice (or Both)

Perception of fairness is not just a cultural thing. People come to the negotiation table with their own belief as to what is fair. Each will have their own set of values and their own need to leave the negotiation with dignity. This underlines the importance of getting the headline agreed (what is the desired outcome) and establishing each party's needs (rather than claims). The first will provide the common focus and goal and the second will shape the eventual deal.

A well-established reputation for fair dealing can be an extraordinary asset. It builds trust and often starts negotiations at a different and more co-operative level.

Sometimes doing a deal that is not fair could be the best long term outcome; for example if the relationship is precious and the concession relatively pain free. Ignoring the fear that this may set a precedent (a common fear expressed in negotiations), the concession may well preserve or even strengthen a relationship and be remembered in future dealings. Or it might just be that it is worth saving the energy and resources of a fight for more important issues.

25.4.1 Trust Is Everything

Trust is a precious commodity. It enables harmony, confidence, and dependability. When broken it causes suspicion, disengagement, and opposition. The dictionary describes trust as 'the firm belief in the reliability or truth or strength of a person; the state of being relied on; a confident expectation'. In business it grows from an assumption of ethical working, of parties sharing the same principles and belief of fairness. So when parties trust each other, there is an assumed common ground upon which to build. Unfamiliar parties will test each other in the early stages of a negotiation to be sure that their assumption of common ground, of trust, is justified.

However, when trust is broken, or even assumed to be broken, it takes a lot of time and patience, and often some risk and vulnerability, to recover. It happens in small increments, each one being tested before the next is considered. In these circumstances it may be wise to change the negotiation team, or individual.

25.5 Relationships

In the long term 'beating' the other side, and causing them harm in the process, may not be a satisfactory outcome.

A good working relationship is one that can cope with differences. Indeed, a healthy organisation will value those differences and work on how they can be used to the benefit the organisation. A good working relationship tends to make it easier to get a good substantive outcome (for both sides). And good substantive outcomes tend to make good relationships even better.

Of course, some people don't care about relationships. They like a quick 'kill' and then move on. It doesn't matter about squashing or bullying the 'opposition' into submission. In fact some people enjoy the reputation of being known to be hard. However, such people eventually run out of others to beat up; their reputation goes before them and better deals can be made without them. The majority of people want a good relationship with those with whom they work, and squashing the opposition doesn't help that!

25.6   The Move Towards Collaborative Working

There is a growing acceptance that better outcomes result from co-operation. In some major infrastructure projects, the UK Government has adopted the principal of collaborative working and has weighted tender analysis towards non-confrontational company policies. Indeed, in 2017 ISO 4401 (based upon BS 11000) was published and it attempts to formalise such an approach, being titled 'Collaborative business relationship management systems – requirements and framework'. It aims to assist companies in establishing and improving collaborative relationships, both within and between organisations. This is an encouraging move and one that recognises co-operation leads to better outcomes.

It is remarkable really that it has been possible to 'standardise' what is essentially behaviour. Of course, it ultimately depends on the willingness of the individual to make it work. It means resisting the instinct to fight or flee, to become defensive and/or seek to blame others when things go wrong, or to reverting to type when challenges occur. However, it does signal a recognition that collaborative working is an accepted, if not ultimately preferred, way of negotiating and carrying out business generally.

25.7   Best Deals

In any negotiation parties need to be clear about the intended outcome (the headline, against which everything else is checked) and each party's needs. Note the term 'needs' not 'claims'. Needs can often be very different to a stated claim. So if needs are known and can be met then a deal is almost certain. Once a party sees that their needs will be met in the emerging deal, their attitude moves towards commitment and co-operation. The best deals emerge from co-operation.

Positional (adversarial) bargaining results in the minimum deal getting agreement. Co-operative negotiation, seeing the situation as an exercise in joint benefit, softens the edges of negotiation and allows outcomes that benefit all parties. The parties are able to seek solutions that enrich a deal rather than pare it to the absolute minimum.

25.8   Staged Resolution

Despite all the above, some issues develop beyond disagreement into disputes. A business would do well to include a staged resolution process in its contracts, starting with face-to-face negotiation through to a third party decision. The following dispute resolution spectrum explains briefly the options that are available.

25.8.1   Dispute Resolution Spectrum

- Collaborative
  - *Negotiation*. The cheapest and quickest way to settle differences. Face-to-face negotiations are the most effective and efficient.
  - *Mediation*. An assisted negotiation, mediation introduces an independent third party to assist parties to find a mutual settlement.

- Non-imposed
  - *Expert opinion.* The use of someone with sector or subject expertise to advise, and possibly recommend, on the facts and possible solution.
  - *Dispute Boards.* Normally three specialists who are available to intervene when problems occur on a project, and to advise on the most suitable resolution process if matters descend into dispute.
- Imposed
  - *Expert determination.* A specialist will investigate a specific matter and make a, usually binding, decision on its outcome.
  - *Adjudication.* An independent third party decides on the outcome of a dispute based upon the relevant contract and law.
  - *Arbitration.* One, or sometimes three, sector experts make a binding decision based upon the relevant contract and law.
  - *Courts.* A judge makes a binding decision based upon the relevant law. This would be limited to an award of money or an injunction to prevent a party taking a proposed action.

25.8.2 An Outbreak of Common Sense

Negotiation aside, of all the above methods of dispute resolution, mediation is the most rewarding and returns common sense to the dispute resolution arena. When a party cannot negotiate a satisfactory outcome to a problem the first thought is usually to involve solicitors. Necessarily, because that is their expertise, they turn a commercial problem into a legal one, and thereafter the language is often not easily understood by the lay person. Time passes, bills are paid and yet solutions rarely appear. Court dates are often months if not years away and the party has no control of the process or the outcome. Mediation allows the parties to a dispute to step off the litigation treadmill and take back control; just for a day (usually). The mediator is there to give the parties the best opportunity to reach a deal. And most do.

- *When to mediate.* The earlier the better, although both parties need to know the issues and positions and to plan a strategy to effectively negotiate a deal. The optimum time must be a point where sufficient is known, and therefore the risk is contained, but the costs are still reasonably low. So probably before full disclosure and witness statements are taken.
- *Who is the mediator?* Most parties feel more comfortable with a mediator from their business sector. Others opt for a lawyer/mediator in the belief that they will know the relevant law. Whilst both may be valid, the best choice of mediator is one who can quickly build a relationship of trust with the parties so that they are prepared to share sensitive information without feeling vulnerable. In mediation the law provides the background but the deal is almost always a commercial one.
- *Who attends?* The wisest parties keep their team lean and keen. The highest level of decision maker, with authority to settle, their legal advisor, and not many others. Experts have usually done their reports and had meetings to agree common ground. It is unlikely that they will agree uncommon ground in the mediation, so it is usually best to have them available on the telephone but not idling their time for a day in mediation.

- *Mediation process.* Usually a day. Typically there will be a main room in which every-one can meet and a room each for the parties. The mediator will usually meet each party privately at the start and then bring everyone together in a joint meeting to agree the issues to be addressed, outline their positions, identify their differences, and agree a strategy that leads to a settlement. When that meeting has served its purpose, the parties will normally return to their rooms and the mediator will shuttle between them, establishing needs and shaping a deal. There is no deal unless all parties agree and most settle, probably because the parties have control and have ownership of the outcome. When they agree the deal is put into writing and the parties sign. Up to that moment they are not committed but, once signed, the settlement becomes an enforceable contract.
- *Mediator's role.* The mediator manages the process, listens, questions (but not like a cross examination), challenges, and encourages. S/he does not advise, impose a solu-tion or pressure the parties into a deal. Even handedness, neutrality, and openness are essential.
- *Party and advisor's roles.* The party's role is to be well prepared, know their walk-away point, and make the best of the opportunity that the mediator offers. The solicitor's and other advisor's roles are to support and advise but also to ensure that their client can make an informed and sustainable decision.
- *Deals with dignity.* The best deals are reached through co-operation – recognising that this is a mutual problem that can be best resolved with a mutual solution. Win-ning should be mutual, although in reality a deal is often shared pain. However, it brings finality and certainty, whereas a trip to court is not only expensive but full of uncertainty. Even the most optimistic lawyer will have a story about the most certain case going the wrong way because the judge got out of bed the wrong way (or out of the wrong bed the right way!). Mediation puts the outcome in the hands of the parties, an outcome that is not imposed by a disinterested third party.
  Another advantage of doing deals in mediations is that the scope of possibility far exceeds that in arbitration or the courts. It can include anything, so long as it is legal, and many deals have included staged payments, work in lieu of payment, dis-counts on future work, even take-overs, and new joint ventures. It is an opportunity to enrich an outcome; an opportunity that is not open to other dispute resolution processes.
  It also allows parties to do deals with dignity. Face is an issue in every culture – losing face is not a happy experience. Mediation allows a party to settle and leave with head held high – face saved and dignity preserved.
- *If it doesn't settle.* Most mediations settle; currently around 80%. So parties entering mediation have a very high chance of settling and getting on with life without the misery of the dispute and continued litigation. The few that do not settle often do so soon afterwards. The momentum and insights gained in the mediation often contin-ues and parties, sometimes with the mediator's continued help, will often negotiate a later deal. However, some parties do not want to take responsibility for deciding a settlement and others may be too stubborn, or too wedded to a matter of principle, to agree a settlement. However, the few that do continue to court often regret the missed opportunity that mediation provided!

25.9   Conclusion

It can be said that justice is best obtained by the parties taking control of their dispute and its outcome, rather than by a judge or arbitrator imposing a decision that can only be based upon the law. So mediation, which has become an established part of the dispute resolution landscape, should be the route for anyone who has failed to negotiate an outcome themselves, and the courts, including arbitration, should be the last resort.

Better still, though, is to deal with an issue before it escalates into a dispute. Face up to it, talk about it, take time to understand and co-operate in finding a solution. It is so much more satisfying and rewarding, and by far the best way to retain relationships rather than risk them being fractured by unmanaged conflict.

26

Dispute Resolution – The Benefits and Risks of Alternative Methods

26.1   Introduction

Virtually all projects have issues during design and construction that require negotiation between the parties and if these negotiations do not resolve an issue after a reasonable period then it will probably turn into a notified dispute.

Disputes can occur in respect of cost, programme, quality, and function, and they can arise on even the most successful projects.

How these disputes are then handled is of critical importance in respect of the costs of running the dispute; the executive time involved; and the ongoing relationships of the parties. Even when an issue does become a notified dispute, negotiations and/or mediation should continue because the parties will be focusing harder on their differences and a settlement may still be possible before formal proceedings are commenced.

In this chapter we will concentrate on the following aspects of dispute management, because they contain a lot of human behaviour considerations.

- Avoiding formal disputes through early communications and negotiations
- Main considerations of the parties when they end up in a formal dispute
- What do commercial clients want out of a formal dispute process?
- Working with lawyers
- Techniques for negotiating settlements.

It is not the purpose of this book to review the mechanics and merits of the various methods of formal and alternative dispute resolution (ADR). In fact, there are so many publications on this subject coming forward all the time that it is hard for even experienced construction industry professionals to keep up, let alone for less experience participants to know where to start.

26.2   Avoiding Formal Disputes Through Early Communications and Negotiations

I am a great believer in doing everything possible to negotiate settlements and this is always my prime objective, even after the disputes have become 'formal' insofar as a notice of arbitration or similar for litigation has been issued.

I have expanded on this approach in Section 26.6, 'Techniques for negotiating settlements'.

Global Construction Success, First Edition. Charles O'Neil.
© 2019 John Wiley & Sons Ltd. Published 2019 by John Wiley & Sons Ltd.

To avoid ending up in a formal dispute that goes the whole distance, communication is the key. It is not easy to get people to sit down and talk objectively once the battle lines have been drawn, but it is still possible if you go about it the right way.

Human emotion is the big obstacle, so getting around that is the challenge. If the senior management of the party in dispute, be it the investor, the contractor, the sub-contractor or the government department, is serious about trying to resolve the dispute before it goes to court or arbitration there are a few simple techniques that are really worth trying:

- Do not leave the handling of the matter solely in the hands of the manager or project director that has been driving it since it started. That person is likely to have too much emotional baggage attached; they also might have created the problem or let it get that far unnecessarily and therefore they will have something to prove and probably be quite defensive of their 'just cause'. In other words, they will have lost their objectivity and commercial realism.
- 'Talk first – write later – and talk, talk again' – this does not necessarily mean you are going soft on the issues in contention, but it does signify that you do not really want to enter into formal proceedings and may be willing to compromise to avoid that.
- Try an informal cup of coffee with your adversary, out of the office of course; work on the relationship if you want to have continuing business with the other party.
- Lay your cards on the table; often adversaries are fighting at cross purposes and have got a screwed-up or inaccurate view of the position of the other side and how they might be willing to compromise. There is nothing wrong with trying a 'without prejudice' discussion.
- The best way to negotiate positively is to 'know your contract' and stick to being objective. You can calmly try to point out to the other party that they are using a wrong interpretation of the contract terms.
- Always keep in mind that success in resolving issues is a direct reflection of the relationships and communications between the parties.
- Stick to the principle that it should not be a dispute, only an 'issue' that needs solving through sensible discussion. Mutual goodwill is enormously valuable of course. That might have evaporated by this time, but that does not rule out compromise as a pragmatic approach.
- Often there will be grey areas in contracts, or maybe even black holes, so you can point out that it is in the interest of both parties to amend the relevant terms of contract rather than spend a fortune trying to resolve the unsolvable.
- Always remember who has the contract with whom – the client and the contractor (but not the client with the contractor's subcontractors or suppliers), or the client and the design consultants; likewise the contractor does not have a contract with the design consultants if they are directly contracted with the client. This can become confusing if the design consultants are novated over to the contractor after the project has commenced, as often happens. So when considering the rights and obligations, risk transfers and claims, or disputes, it is important to keep in mind which specific parties have direct contracts with each other.
- A three-way arrangement whereby the client has a Contract with the contractor and separate contract with the principle design consultants will only work smoothly if the communications protocol and management platform work with discipline and efficiency, otherwise this arrangement can be difficult and cause lots of problems.

26.3 Main Considerations of the Parties When They End Up in a Formal Dispute

When all else fails commercial managers want a dispute resolution process that:

- Is as fast as possible, without unnecessary delay
- Minimises legal fees and executive time and disruption
- Has confidentiality if possible, e.g. arbitration
- Minimises the damage to relationships; ongoing business between parties might be a very important consideration, but so is the maintenance of personal relationships, because it is a remarkably small community in most industries and businesses and people move between companies fairly regularly these days.
- At this stage everyone wants to win, obviously, but realistically there are no guarantees of this in dispute resolution – so if the result goes against you then you want to be satisfied that you have had a fair and reasonable hearing and that the decision handed down is legally and contractually correct, logical and well reasoned.

26.4 What Do Commercial Clients Want Out of a Formal Dispute Process?

This might seem like a simplistic question but it will probably help us decide which dispute resolution process is the most suitable one to use to determine a particular dispute.

The reasons that companies enter into dispute can be many and varied and they have an important bearing on how a dispute should be managed.

I categorise these reasons as being either 'tactical' or 'genuine'.

There are five reasons that I classify as being tactical:

- To enable an independent party to decide the issue, because the negotiators are not authorised or prepared to make a decision.
- To encourage settlement, because the other party won't talk.
- To create leverage on the other party for some related reason, financial or company political, e.g. a trade-off position.
- To defend the ego or incompetence of a manager or a previous management decision
- To delay payment.

Tactical reasons can be very frustrating, time consuming, and expensive, but it is a fact of life and human nature that they occur, and when they do we need to be best equipped to deal with them.

However, there are also some genuine reasons why people fail to settle their differences and end up in formal proceedings and most of them will fall under one of the following two categories:

- A genuinely perceived difference in contract interpretation
- Differences related to the value or timing of payment for goods or services.

Summary: keep talking and trying to settle right up to the door of the courtroom.

> Remember that whenever there is a problem with a project it can invariably be traced back to a breakdown in communication somewhere.

26.5   Working with Lawyers

In September 2008, I was privileged to be invited to be a guest speaker by the NSW Law Society in Sydney, in my capacity as executive director of Bilfinger Project Investments in Australia and as a chartered arbitrator. I was probably the only non-lawyer in the room, with an excellent attendance of more than 130 people.

The topic was 'Arbitration from a commercial client's perspective', but I took the opportunity to broaden the topic considerably and included quite a bit about what the management of major companies expect from their legal advisors. I prepared the speech very carefully and circulated it to four legal and commercial colleagues in advance for checking, as I wanted to be frank and constructive but not offensive. My colleagues liked the approach, helped me amend it in a few places, and I think spread the word; hence the good attendance.

In summary, my main point in respect to 'working with lawyers' is that the lawyers should not be left to run the show entirely as they see fit. The client's commercial manager should be working hand-in-glove with the lead lawyer on the case, but amazingly this is not always the situation, and client's then wonder later why the case went off the rails and the fees were enormous, often out of proportion to the value of the case.

It seems obvious to me, but the lawyer/commercial manager partnership is the only way to go in a complex commercial dispute, irrespective of whether it will be resolved by arbitration, litigation, adjudication, expert determination, negotiated settlement, mediation, or other form of dispute resolution.

The contributions that the lawyer and commercial manager each make are complementary and essential in order to get the optimum result:

- Lawyer
  - Legal and case knowledge
  - Legal strategy
  - Proper interpretation of the contract
  - Separate the emotion from the facts
  - Assess risk
  - Objective, cost effective advice.
- Commercial manager
  - Status of the issues, claims and disputes
  - Facts and figures
  - Corporate objectives
  - Status of relationships
  - Implementation of agreed strategy.

One point that I did make to my audience of lawyers is that the biggest mistakes they can make are to (i) give 'sitting-on-the-fence' advice; and (ii) string cases out. They will only do it once because they will most probably lose that client altogether when the dust settles and the client realises what has occurred.

Also, it is not uncommon for lawyers who have been involved from the start to give advice that suits their client's commercial objectives and not a hard-nosed objective interpretation of the contract. This 'slanted' advice can prove expensive, both for fees and time wasted. It can actually happen subconsciously to a degree – again, the human factor.

It is important to remember that big busy corporations like to slot situations into boxes, make provisions in the accounts and get on with life. In most cases they would like the dispute resolved as soon and as inexpensively as possible.

Not always of course, as happened with Seven Network Ltd versus News Limited, one of the longest and most expensive cases in Australian history. During the course of the trial the judge questioned the wisdom of the claimant continuing to pursue the case and stated that it was 'extraordinarily wasteful' and 'bordering on the scandalous' and led him to caution against appealing the decision, which did not deter the claimant, who appealed and lost, producing a further judgement running to 359 pages. It seems that senior management lost their perspective well and truly.

http://www.australiancompetitionlaw.org/cases/c7.html

My speech: 'Arbitration from a Commercial Client's Perspective' was published in full by the Chartered Institute of Arbitrators in the UK (CIArb Journal Feb 2009).

A client claimed against a small legal firm for excessive charges over a five year period of litigation. Following mediation and the threat of disciplinary proceedings, the legal firm refunded $400 000 fees and suddenly found it was able settle the case and wrap it up within a further 12 weeks, with no further legal fees being charged. The client was an individual, not a corporation, and it was clear the legal firm had been both overcharging and stringing the case out, which was even worse because the client was 92 years old.

26.6   Techniques for Negotiating Settlements

Negotiating settlements is an area in which I have specialised for many years as an independent negotiator and for which I have received appointments from government authorities and corporate entities.

My philosophy and firm policy has always been to use all possible avenues to achieve fair and reasonable settlement of claims and disputes through negotiation, rather than through the courts, arbitration, adjudication, and expert determination. My approach is different to mediation and conciliation in that it is much more proactive. It has proven to be very successful.

As an independent settlements negotiator I have come across many situations where clients and contractors are at loggerheads and communications between them are such that the only way forward appears to be through formal dispute proceedings, whereas a calm analytical approach with a bigger picture perspective can very often resolve differences and arguments over claims.

When serious differences arise between the parties over claims or other contractual matters, it is best to never call them disputes unless they actually end up in formal dispute proceedings, because the word 'dispute' has a negative connotation that sets the wrong scene for a negotiation.

To me they are just issues and there is nearly always a way to resolve them through negotiation. An interesting philosophy to adopt is to say to yourself 'this issue will be resolved by five years from now one way or another, so let's find a way for the parties to resolve it now' and then work patiently with them to find that way.

This sort of negotiation is more than a traditional mediation in that the negotiation process is very proactively led by the independent negotiator, but first you have to win the confidence of the parties that you are experienced in the subject under debate and that you will be objective, completely impartial, and fair and reasonable in the advice you put forward.

Even when the parties have reached an apparent stalemate, ceased communicating and have resorted to firing bullets at one another through their lawyers it is not too late to negotiate, although it may be difficult to get them in the same room to start talking.

A typical example of this sort of situation is described at the end of this chapter in 'Paddy's Market, Haymarket, Sydney, – A classic failure to communicate'.

As another example, in one project I acted as the independent negotiator and chaired settlement meetings on more than 100 disputed construction and extensions of time (EOT) claims, many of which were extremely complex, but we agreed settlement on 90% of them. The negotiations took 18 months and some of the individual claims took several months. The parties, who were from three different countries, were keen to avoid long and costly international arbitration so there was a willingness to compromise, which is an extremely important factor for success. The process used was far more proactive than mediation. I was requested by the employer, the contractor, and the project manager to use my arbitration experience to analyse each claim and advise my view of the likely outcome if the claim ended up in arbitration. The contract documents were complex and the terms difficult to understand, seeming ambiguous and contradictory in places until you flowcharted the rights and obligations taking into account precedence of documents. The main problem with the contract was that it had been amended twice in the previous four years, each time very badly. It was surprising that the legal advisers to each of the parties had agreed to those amendments.

I discovered at the outset that the parties had widely different views on the interpretation of the heavily amended FIDIC contract, so after realising this I put on my arbitrator's hat and wrote an 'Interpretation and Application of the Contract', including sufficient reference to the rights and obligations of the parties to cover the first batch of claims that were under debate.

So armed with this, we then had further meetings at which the parties presented their cases in detail and I gave my view on what I considered to be the correct interpretation of the contract, i.e. the likely result if it went to arbitration. With each one I explained my reasons in detail, going through the contractual obligations step by step. For quite a number of the disputes, after I gave my view the parties then took some time to consider the matter further or talk to their lawyers and on most of the claims one party or the other came back with a reasonable settlement offer. Where this didn't happen the first time around we tried again, and if still not successful we then parked the claim for a few weeks; almost every time we did this we reached agreement in the third discussion.

During the overall course of the negotiations, I wrote the following 'Expert Opinions' or indications of likely arbitration findings and issued them to the parties to help them reach settlements:

- *An Opinion on the Interpretation and Application of the Contract Agreement and on the Rights and Obligations of the Parties.*
- *On the Meaning and Application of the Terms of Contract relating to Re-Design by the Contractor.*

- *On the Employer's Entitlement to Reduce the Scope of Work of the Contractor and directly employ and manage subcontractors simultaneously with the Contractor's Works and the cost implications thereof (covering Contract Works, Nominated Subcontractors and Provisional Sums).*
- *On the Contractor's Entitlement to Loss of Profit and Overheads on Omitted Works.*
- *On the Employer's Entitlement to Deduct Counter Claims from the Contractor's Account under the terms of the Contract.*
- *On the Meaning and Application of Amended Terms of Contract relating to Claims for Extensions of Time and Applicable Costs and the respective Contractual Processes.*

I have listed the above examples to demonstrate the importance of really understanding your contract. Put the time in to do this and get legal advice in doing so; it will save you and your company a great deal of time and cost in managing your claims and disputes, let alone a lot of frustration and sleepless nights.

Paddy's Market, Haymarket, Sydney – A Classic Failure to Communicate

The markets had been operating for 150 years and accommodate more than 800 stalls. The markets are managed by a government authority that works in conjunction with the Stallholder's Association. In 1987 a property developer obtained the redevelopment rights and a long-term lease to the site from the government, with a view to putting a 35 level multi-storey building on top of reconfigured markets and a podium of conventional retail and cinemas.

The new architectural plans were submitted for planning approval and put on public display, but up to this time the developers and their design team had not met and consulted with the Stallholder's Association at all. The stallholders were incensed because the plans took no account of the original stall and product locations, the people and goods flow or the loading docks. Of particular concern were the stall locations and sizes, because these related to their leasehold goodwill value. Typically, the values of the stalls ranged from AUD100 000 to AUD250 000 at that time and were real assets.

The developers had also agreed with the government to temporarily relocate the markets for three years to some disused railway sheds 2 km away, again without any consultation with the stallholders.

The Stallholder's Association responded with court injunctions to prevent the planning approval, the temporary move and the development as a whole. The injunctions resulted in a stalemate for 18 months, at which time I was asked by the developers to act as mediator. The stallholders were suspicious of my role, but agreed to a meeting of all parties, with two representatives each. At the meeting I quickly confirmed that there still had not been any contact between the developers and the stallholders, except through their lawyers. The first meeting was fiery and the stallholders walked out after 10 minutes. The second meeting one week later was similar, but I was able to exchange personal phone numbers with the Stallholder's Association president. We then met informally over some lunch, following which I had some useful meetings with the Stallholder's Association committee at which they explained their concerns. From this I was able to convince the developers that their plans were completely impractical for a number of reasons. They rebriefed their architects and design team to start again, this time working in close consultation with the stallholders. After four months everyone agreed on the redesign and the logistics for the temporary relocation. I attended most of the initial redesign meetings to make sure that everyone was communicating and staying on track.

The stallholders cancelled their legal proceedings and planning approval was granted. At this stage the developers sold out to another property company. The lengthy delay had burdened them with too much holding cost. The new owners simplified the podium and raised the multi-storey by several levels and eventually completed a successful development.

The risk management lesson here is very obvious – a little communication goes a long way. It is the easiest form of risk management, but breakdowns in communication probably cause more problems than anything else.

Happy ending: I was guest of honour at the Stallholder's Association next Christmas dinner, with around 2000 guests at the party.

27

Peer Reviews and Independent Auditing of Construction Projects

In this chapter we examine:

- Why are peer reviews and independent auditing needed? Are they cost effective?
- Difficulties encountered in having independent reviews.
- The value at different stages of a project.

Peer reviews and independent project audits are a very cost effective and prudent form of risk management because:

- They are broader based and properly independent compared to internal audits.
- They can help minimise the risk of poorly conceived projects getting off the ground.
- They may provide early warning of potential issues that might not have been recognised internally.
- They reduce the risk of cost and programme over-runs and disputes through identification and early warning of project issues.
- They can provide an opinion on all claims and disputes as well as operational matters.
- They can provide a level of sophisticated review not necessarily available on major overseas projects.

The review team should work closely with site management, and corporate management if necessary, with the prime objective being to prevent projects running into trouble.

The real value of independent reviews comes from starting them early in the project life and conducting them at least every four months, reviewing all aspects of the contract, with recommendations made for corrective actions to mitigate if problems are found. It is most important to follow up and build on previous audits.

The scope should be agreed with the client, including all basic areas that might put the project at risk:

- Health and Safety procedures and standards
- Design development and cost planning
- Progress versus programme; expediting measures
- Drawdowns versus budget and programme
- Cash flow management and financial reporting
- Costs-to-complete versus programme

Global Construction Success, First Edition. Charles O'Neil.
© 2019 John Wiley & Sons Ltd. Published 2019 by John Wiley & Sons Ltd.

- Project reporting
- Site resource capacity and capability, including key people suitability, staffing levels, consultants, plant and equipment
- People management, team spirit and employee satisfaction
- Subcontractor status and management, including their payments
- Compliance with contractual obligations
- Specification compliance
- Planning, building and authority approvals
- Issues at large, claims and disputes
- Communications and relationship management
- Meetings and documentation discipline
- Environmental requirements
- Quality assurance, including NCR's (nonconformance reports) and RFI (request for information) registers
- Corporate governance
- Completion and commissioning plans
- Staff compliance with the 'first to know' rule for the project director, especially with *'early warnings'*.

The team should start with design development and cost planning as this combined process is the most critical phase of a D&C project; when you make or break the project. These two processes must be run in parallel. It is pointless having the design consultants develop the design unless there is constant cost checking against the budget or Bid price. This is when independent peer reviews are most valuable. They have proven to be very effective and valuable in the design process.

After a successful design development and cost plan phase it should then be a matter of efficient head contract and subcontract management and tight programme control, subject to the usual types of variations and standard delays such as weather and approvals, etc. However, this is oversimplifying it, especially with Public Private Partnership (PPP) contracts where project managers commonly don't understand the obligations of the different parties, and the implications and liabilities arising from the contract, and therefore they often don't realise they are getting into trouble.

All of this is based on two major assumptions of course:

- That the estimators got it right in the first place and the directors have not chopped too much out to win the Bid.
- That the Bid passed all the risk management checkpoints and red gate reviews and was thoroughly checked for potential risks.

Major construction companies generally say they have these processes in place with their internal audit team and therefore don't need external audits, but they still have projects that get into trouble.

However, the problem is that internal audits tend to:

- Not look at the bigger picture
- Check historical figures
- Not critically examine forward programmes, resources and costs-to-complete
- Have loyalty or conflict of interest issues
- Not be completely independent and objective.

It is not always easy to convince construction managers that they benefit either way from peer reviews and external audits. The reviews and audits should be seen as 'safety first' low cost risk management. If explained properly to project directors they should be welcomed, because it becomes a *'win–win'* situation for them, as detailed below.

- The upside is confirmation that everything is healthy and on track.
- Alternatively, if the external audit team unearths some problems and they are corrected or mitigated before they get out of hand then the project director still gets the credit.
- If an innovative solution suggested by a design review saves time and cost, the project team still gets the credit.
- The fees are highly cost effective in proportion to project value.

Professional pride and egos can be major barriers to having peer reviews or independent audits accepted by senior corporate or on-site managers, but they should remember that *'it's too late to try and shut the gate after the horse has bolted'*. The question is how to convince headstrong, stubborn project managers that they should use external assistance to identify, capture, track, and mitigate risk.

Stakeholders in a project that can benefit from independent audits and peer reviews are:

- The client, with an overview of the entire project
- The head contractor
- Major subcontractors
- Principal consultants
- Any combination of the above.

The independent audit team must have extensive experience in the following areas, or in specific project disciplines as required.

- A wide range of major projects
- Design development/cost planning
- International forms of contract
- Contract management and cost control
- Performance programming and expediting
- Operations and asset management
- QS (quantity surveying) processes
- ADR (alternative dispute resolution).

So what point of difference do independent audits really offer in respect of risk management?

- Team members have the experience to 'smell' issues in the making and provide early warning.
- They will make their own assessment – irrespective of reports and programmes from the project team.
- The approach is hard-nosed and objective on current status combined with a strategic look ahead.

Starting early is the key to success and the cost is really minimal.

Section F – Emerging Conclusions

28

Conclusions and Recommendations

28.1 Overview

Our authors have examined the key areas that we consider have high importance in any assessment of the industry, commencing with:

- The state of the industry globally
- Causes of project and corporate failure
- Consultants' challenges in the international market
- Issues with supply chains.

We then provided structures, explanations, and details on the important elements of the main themes that we consider essential for successful project outcomes:

- People and teamwork – management competency, structuring projects, personnel issues.
- The right framework – business models, forms of contract, collaboration, and Public Private Partnerships (PPPs).
- Management of risk – human factors, effective processes, technology, successful projects.
- Robust processes – for corporate and project management, conflict management, audits.

It is clear from our reviews that such a global and diversified industry will always attract a wide range of issues (and opinions). It is also clear that many of today's issues are not really 'new' and that the industry has faced many of the same core challenges for decades. Some common and extremely important themes have emerged from our chapters and we summarise these as follows.

It has become evident from our findings that some sections of the global construction industry are currently doing well, but many participants are not. Also, that some countries have a much healthier construction industry than others. **At a high level, one can conclude that:**

- There are still far too many project failures around the world, with high cost over-runs, delayed completions, and poor quality.
- There are too many contractor insolvencies (for a number of different reasons).

- Clients will nearly always achieve better project delivery with less risk if they enter into contracts that suit the specific project and the mutual relationship that they want with the contractor and/or consultant.
- Contractors, consultants, and clients need to:
  - Invest in and empower people in their organisations, with more extensive training and with a deeper understanding of the impact of human dynamics
  - Implement robust business processes
  - Place far more importance on risk management
  - Embrace technology
  - Give much greater recognition to the criticality and viability of the supply chain.

We have observed that contractors world-wide appear to be operating on two distinctly different levels of efficiency, which are summarised in Table 28.1.

Table 28.1 Different Levels of Contractor Efficiency.

Progressive and well-managed contractors that have:	Contractors who are persisting with 'old style' industry practices and cultures and who:
Tight financial controls	Are surviving from day to day using unprofitable business and contracting models
Robust strategies and policies governing bidding, margins, and liquidity	Have management who are living in the past and have limited competence for the present (never mind the future)
Effective processes, including stringent risk management, and who can often make use of new digital platforms to obtain efficiency gains	Have inadequate risk management, ineffective control, and reporting processes, and are seemingly unable to cope with new technologies
An understanding of the human factors in all areas and an ability to embrace these	Do not understand the human dynamics
Significant reinvestment in R&D, training and technology	Do not invest sufficiently in R&D, training and technology

No amount of legislative change can protect 'old fashioned' contractors from their own inefficiencies and out-of-date management approaches and practices. Lenders and shareholders might put on a lot of pressure, but ultimately contractors like these will not survive in the long term if they are unable to 'modernise'.

There is no doubt that construction is a tradition-bound industry, with methods and processes being passed down from one generation to the next as being 'the only way to do it', but these traditional methods are no longer adequate for survival when you are competing against really progressive contractors who are running their operations in a hard-headed business manner and using very efficient methods and new technologies to manage their finances, personnel, general administration, risk assessment, project delivery, etc.

28.2 Where Is the Global Industry Headed?

To the industry outsider, it seems quite ridiculous that the same construction sector problems have been recurring in cycles for decades and keep happening with each new generation of managers – despite (or sometimes, because of) endless independent and government investigations and industry body reports. This is unacceptable. Construction is such an important industry in every country in terms of employment and GDP share that it is way past time that real changes should take place in order to move on from obsolete and unprofitable business models and practices, and in some countries, ongoing corruption.

In such a low margin industry it does not take many 'bad' projects to tip a construction contractor over the edge and into bankruptcy, which is then followed by major problems for employees and the supply chain and losses for the investors and banks and also for the economy as a whole.

However, it is too easy to blame over-competitiveness and low margins for these failed projects or contractors, particularly because the majority of contractors are profitable enough to survive year in, year out, even if they are not really prospering, and a minority of contractors and their supply chain are actually doing very well.

Not all clients are perfect either, with some having unrealistic expectations coupled with inadequate knowledge of the contracting process, often entering into unsuitable forms of contract; others can be just plain inefficient, and worst of all are those that turn out to be untrustworthy bad payers.

Risk management plays a big part in managing for success and it is really all about people. Organisations need to invest in the best and most appropriate corporate processes available, but at the end of the day the success or failure of the project will depend on all stakeholders having the right people in the right positions, efficiently using these right processes.

It is crucial to make sure the 'lessons learnt' from both successful projects and disastrous ones are passed on to the next project and the next generation. One of the main objectives of this book has been to present these experiences in an easy-to-read, concise, and practical form for all participants in the industry and, in particular, we hope that young professionals can take them on board and not learn them the hard way. Managing for success is the over-riding goal.

It is time that CEOs, directors, and senior managers of many construction companies moved on from the industry's embedded cultural practices and started to run their organisations in a more robust business manner (for example, stop the 'must win' chase for projects).

At the same time, it is important that they invest sufficiently in training and new technology in all the areas necessary to achieve real and sustainable corporate and project efficiency.

28.3 Key Observations and Recommended Actions

Listed below are the key observations that we have made from our review of the industry and our recommended actions to assist department heads, CEOs, directors, and senior managers in making the necessary step changes that will make construction a more stable industry. It is essential that clear leadership needs to be provided by both contractors and clients towards improving the efficiency of the industry.

28.3.1   Legislative Change

This will vary from country to country, but it is required to improve and ensure high industry standards and to protect the industry from unscrupulous and/or really inefficient clients and contractors in areas such as:

- More stringent financial reporting and auditing
- Strict adherence to health, safety, and environmental standards
- Security of retentions and payments
- Pension fund protection
- Corporate social responsibility (CSR)
- Provision for greater accountability by senior managers and directors.

Government departments and politicians responsible for such changes would be well advised to be more proactive in comparing notes with other countries and listening more carefully to their industry organisations, rather than wait until disasters such as Carillion or Grenfell Towers in the UK occur before taking action. The signals for problem areas are normally loud and clear.

28.3.2   Client Contractor Relationships

Adversarial relationships are self-destructive. This is well recognised by successful contractors and experienced clients. Good relationships produce good projects and good projects produce successful contractors.

There are some fundamental requirements and principles that are required in order to develop strong client contractor relationships:

- Mutual recognition of the desired result of each other, i.e. a quality project, on time, and on budget for the client, and a profitable, trouble-free construction process for the contractor.
- Mutual respect and awareness of business, personnel, and cultural values.
- Proactive implementation at the outset of effective communications and relationship management protocols.

The following are our additional key observations and the recommendations we put forward to address these challenges:

1) **The supply chain has been taken advantage of (in numerous different ways) by clients, contractors, and consultants for too long.**
   - This must change because all supply chain participants comprise an integral part of the entire industry and are heavily relied upon by contractors and clients. We must seek to find more and better 'win–win' outcomes for project delivery.
2) **Competency of people, team building, and recognition of the importance of human factors are constantly recurring themes in corporate and project management.**
   - Corporate and government senior management competency is a fundamental thread throughout this book, because at the end of the day all good (and bad) management practices and cultures emanate and are led from the top.

- The importance of having capable teams, good team spirit, and of understanding human behaviour cannot be over-emphasised. In return, employees expect openness, honesty, and trusting delegation of responsibilities. There is no room at the top for dominating, noncommunicating, incompetent managers (who may only have got to that level in an organisation through cronyism or time serving).

3) **Real improvements have occurred with several forms of direct construction contracting and business models.**
   - Internationally, the 2017 FIDIC Suite of Contracts and the 2017 New Engineering Contract (NEC4) represent substantive improvements in standardised contracts that enable efficient management of major projects, and provide for proactive collaboration and non-confrontational resolution of conflicts and disputes.
   - Similar improvements have been made in several countries with their standard forms of contracts and globally there is now a good range of different contracting formats available to suit specific types of projects.

4) **PPPs, despite having been around in various forms for a long time, still require significant review today** of contract terms, risk share, and basic understanding to ensure that these projects are attractive to clients, consultants, potential investors, contractors, and facilities management services companies.
   - The basic PPP format has only seen incremental changes in recent years (up until the publication of this book). Because PPPs are growing in importance worldwide, especially in developing countries, the changes outlined in Chapter 15 should be addressed urgently to ensure that the full lessons of the past are utilised for a successful future.
   - The following are of particular note for new PPP markets in this regard:
     - The current forms of PPP contract do not always reflect a level playing field, with the 'allocation of risks to the party best able to manage them' and this is driving potential participants away, to the overall economic detriment of potential new projects.
     - Excessively high Bid costs are continuing to make PPPs a less attractive and more risky investment for the private sector and for contractors (and consultants).

   In a number of countries there is far too much corruption and diverting of PPP investment funds, so lenders, investors, and contractors need to tighten up the supervisory governance substantially, as this is also driving away potential participants.

5) **Successful clients acknowledge that cheapest is not always best and can be risky.**
   - These clients are prepared to negotiate fair and reasonable contract terms on the basis that they require a quality project delivered on time and on budget, with minimal cost variations and disputes. To achieve this, they are prepared to work closely with contractors and consultants in the planning, design, and budgeting in the lead-up to signing the construction contract.
   - They will then enter into an appropriate form of contract to suit the project, such as ECI (early contractor involvement), or one that might include some form of risk/reward sharing, or it might be an alliancing contract.
   - There are several contracting business models available, but all of them will only work well if clients, including governments, recognise that it is in their best interest if contractors and consultants are financially sound and have positive cash flows.

6) **Beware of clients who portray themselves as being 'smart and tough'.**
   - Typically this is the sort of client who likes to negotiate the price down to a level that is actually unrealistic for the delivery time and quality. They are then prepared to argue and fight all the way through the contract delivery.
   - This type of client is also often the bad payer; on time at the start, then getting slower as the work progresses, and finally holding out on the completion payments.
   - Property developers, who are often highly geared, have a particularly bad reputation in many countries as bad payers (although 'enlightened' developers do work closely with their team, nearly always resulting in a better outcome for all the parties – and repeat business)

7) **Risk management is being poorly implemented by many clients, contractors, and consultants.**
   - The value of good risk management appears to be badly misunderstood and under-estimated by many. They are consequently not making best use of the available processes.
   - They often do not seem to equate effective risk management with ensuring project and corporate financial security and profitability.
   - Chapter 18 describes the risk management processes used by Kiewit Development Company in the US and Canada, which provide a good core model for all contractors.

8) **The industry must embrace new technology.**
   - Significant change is always difficult in any industry, and implementing new digital technologies as aids to design and construction certainly requires strong motivation and support from senior management, because:
     - Construction is a particularly traditional industry, long resistant to change and innovation.
     - There may be significant capital expenditure required to incorporate change.
     - New specialised employees may have to be recruited.
     - There has to be a commitment to employee training at project and corporate level
     - Advances in technology are occurring rapidly and some organisations are too slow to absorb and adapt.
     - The use of digital technology is expected to grow rapidly and will define the difference between progressive and innovative managers and 'the old school'.

9) **Robust processes for corporate and project management.**
   - Clients, contractors, and the supply chain use either proprietary, in-house or a mix of both for data gathering, control, and reporting systems.
   - Whichever process that is used it must be based on up-to-date and accurate data and it is essential that the real-time control and reporting formats and systems are easy and simple for employees to use.
   - The process must be able to provide this real-time data for all essential management components, such as cost, financial status, progress, risk management, and corporate governance updates. Strict discipline is required in reporting.
   - It is not acceptable to be announcing project and corporate figures and progress reports that are even three months old, let alone six months out of date.

- Bad news travels fast. It is completely unacceptable for contractors to suddenly announce that they have just become aware of losses on a project that has been under construction for two or three years, as we have witnessed in various parts of the world in recent years. This is senior management incompetency at its absolute worst!

28.4 Final Thoughts

There are signs that the continually increasing global demand for infrastructure and the readily available finance from governments, PPP investors, and the private sector will further push the demand for the construction industry to expand. On balance, the global industry should be poised to go on to a better and brighter future, with large volumes of work available in infrastructures, residential, industrial, and commercial.

However, to succeed legislative changes are necessary and the industry itself must make the badly needed changes that we have identified in order to achieve greater stability and profitability. If this is not done then the industry will fail to meet expectations.

The messages and recommendations made in this book are aimed first and foremost at the senior leaders in government, client organisations, contracting companies, and construction industry organisations, because real change can only be achieved by all of them working collaboratively and with a sense of urgency. And then firmly advising politicians on the requirements.

For far too long we have been hearing about the need for these changes, but they have been too slow coming so it is time for some meaningful and ongoing leadership to give the construction industry the viable platform across all sectors that it should have as one of the world's most important industries.

Failing to plan and adapt is planning to fail – it's your choice. Good luck.

Charles O'Neil on behalf of our team of authors, 2018.

Appendix A

Source: Reprinted with permission from Building: 2nd March 2018

Agenda

Comment / Second opinion

Take me to your leaders

The absence of leadership in construction is deafening, says Rudi Klein. Who will step up to drive through the industry's urgent agenda for change?

Within a period of 12 months the image of construction has been battered to a pulp. Yet there isn't a coherent voice coming from the industry that is providing leadership and a way out of the crisis.

In February last year Professor John Cole's report on his inquiry into the dangerously defective masonry in 17 Edinburgh schools condemned the lack of competence and accountability in the industry. In December Dame Judith Hackitt's interim report on her inquiry into the regulatory framework governing building safety deplored the lack of an integrated delivery process.

And to cap it all, we had the Carillion collapse revealing to government and public at large the dirty underbelly of construction. Many have compared Carillion's operations to a Ponzi scheme - robbing Peter to pay Paul. In fact Carillion had perfected the art of holding onto the cash of its supply chain. Firms paying "commissions" to get work. Others being sucked in by being given loads of work and then put out of business because they were paid either late or not at all.

Firms in the supply chain had no idea who were the ultimate masters or clients. Neither did clients know who was in the supply chain. For example, G4S would contract government work to Carillion, which would then subcontract it to another firm - and this subcontracting would continue on. UK construction has become the outsourcing capital of global construction. And we wonder why UK construction is so costly.

Carillion - like many other large companies - was an empty shell or facade. It did not have any connection to a product. Therefore it was not held in check by any ethos that subscribed to professionalism through enhancing the product for the long-term benefit of its clients. Instead it marketed itself, laced with plenty of management-speak, as promoting sustainable and value for money outcomes - all code for cheapest price. Lack of assets or resources did not matter. Jarvis was doing the same thing years ago, and look what happened to them. As an

 Over the past few weeks I have been asked many times: why did public sector procurers select Carillion and other companies like it? Answer: many construction procurers do inertia better than anybody else

aside, the bidding process in the nuclear industry prioritises the ability to adhere to quality and programme before price.

The Carillion business model - using the supply chains as a receptacle for risk dumping - was contagious. In order to compete, many big construction and property companies have adopted this approach and are still using it. It needs to be dumped.

Over the past few weeks I have been asked many times: why did public sector procurers select Carillion and other companies like it? Answer: many construction procurers do inertia better than anybody else. They lack sufficient knowledge and experience to challenge what is presented to them. Often they are incapable of assessing the quality of the engineering input which, these days, forms the largest value element of construction and services delivery.

But where is the leadership in the industry to give a response to all of this? Answer: nowhere. Yet there is an agenda that demands change. Even the government seems to have woken up to this. Here is Greg Clark, the business secretary, speaking in the House of Commons on 30 January: "The lessons and the scrutiny of what went wrong in Carillion [...] and in the oversight that took place across the whole of the public

sector in terms of contracting, need to be looked at and will be looked at [...] Whatever actions are required from that, we will take." So, what changes are needed?

Firstly, we need radical change in procurement practices to embrace alliancing and also to allow for more SME participation. The majority of projects never come in at the price that was agreed at the outset, because cost plans are never underpinned by any robust risk management process. The whole delivery team needs to subscribe to that process. But, in spite of advice to this effect in numerous reports and initiatives over the years, it rarely does. While encouraging the industry to take up digital and manufacturing technologies it is vital that we now prioritise addressing the dysfunctionalities in the delivery process.

Secondly, we need radical improvements in payment security so that all payments for construction and services delivery are protected. Public sector procurers should now insist that the payment process commences much earlier to cover offsite activity such as design and assembly; this should be required all along the supply chain.

Thirdly, we need radical improvement in quality and standards through licensing reputable firms that can demonstrate the appropriate technical capabilities. Nearly every state in the US has a licensing system for construction works. Queensland, Australia, also has such a system, but it additionally requires firms to have a certain level of assets, which is a requirement too far - just imagine introducing that here: we could lose thousands of firms.

Finally, we need radical improvements in responsibility and accountability for what we deliver. This should embrace effective enforcement of regulations and standards.

Here is my challenge. Those with a drive and energy to take up this agenda should now come forward. Let's not wait for another image-sapping disaster.

Rudi Klein is a barrister and chief executive of the Specialist Engineering Contractors' Group

Index

Global Construction Success, First Edition. Charles O'Neil.
© 2019 John Wiley & Sons Ltd. Published 2019 by John Wiley & Sons Ltd.